Lecture Notes in Electrical Engineering

Volume 581

The book series *Lecture Notes in Electrical Engineering* (LNEE) publishes the latest developments in Electrical Engineering - quickly, informally and in high quality. While original research reported in proceedings and monographs has traditionally formed the core of LNEE, we also encourage authors to submit books devoted to supporting student education and professional training in the various fields and applications areas of electrical engineering. The series cover classical and emerging topics concerning:

- Communication Engineering, Information Theory and Networks
- Electronics Engineering and Microelectronics
- Signal, Image and Speech Processing
- Wireless and Mobile Communication
- Circuits and Systems
- Energy Systems, Power Electronics and Electrical Machines
- Electro-optical Engineering
- Instrumentation Engineering
- Avionics Engineering
- Control Systems
- Internet-of-Things and Cybersecurity
- Biomedical Devices, MEMS and NEMS

For general information about this book series, comments or suggestions, please contact leontina.dicecco@springer.com.

To submit a proposal or request further information, please contact the Publishing Editor in your country:

**** Indexing: The books of this series are submitted to ISI Proceedings, EI-Compendex, SCOPUS, MetaPress, Web of Science and Springerlink ****

More information about this series at http://www.springer.com/series/7818

C. Shreesha · Ravindra D. Gudi
Editors

Control Instrumentation Systems

Proceedings of CISCON 2018

Springer

Editors
C. Shreesha
Department of Instrumentation and Control Engineering
Manipal Academy of Higher Education
Manipal, Karnataka, India

Ravindra D. Gudi
Department of Chemical Engineering
Indian Institute of Technology Bombay
Mumbai, Maharashtra, India

ISSN 1876-1100 ISSN 1876-1119 (electronic)
Lecture Notes in Electrical Engineering
ISBN 978-981-13-9418-8 ISBN 978-981-13-9419-5 (eBook)
https://doi.org/10.1007/978-981-13-9419-5

This Springer imprint is published by the registered company Springer Nature Singapore Pte Ltd.
The registered company address is: 152 Beach Road, #21-01/04 Gateway East, Singapore 189721, Singapore

Preface

Control Instrumentation System Conference (CISCON) is the annual conference event organized by the Department of Instrumentation and Control Engineering, Manipal Institute of Technology. The Department initiated CISCON in the year 2004 to provide a platform for its first batch of B.E. in Instrumentation and Control Engineering students to have an interaction and exchange of ideas with their counterparts in and outside the institution. This is first of its kind in the institute and under the able leadership of Dr. V. I. George. With very few institutes in the country offering this specialized interdisciplinary course, people working in both instrumentation and control engineering sought after for this conference every year and this has gained lots of recognition. The conference has been sponsored by national research organizations like Defense Research and Development Organization (DRDO), Board of Research in Nuclear Sciences (BRNS), Indian Space Research Organization (ISRO), and Council of Scientific and Industrial Research (CSIR), to name a few.

The proceedings of CISCON has been brought out regularly since its inception. In 2015, it was decided to bring out the published papers in Scopus-indexed journals to give additional incentive to the authors who put forward their research articles to CISCON, and the same trend has continued till 2017 with a rapid increase in submission. This year, the organizers have decided to publish the presented/accepted papers as part of Lecture Notes in Electrical Engineering published by Springer Nature to add further value to the publications. The conference has attracted a large number of papers in varied disciplines like process control, automation, renewable energy, robotics, image processing, sensor, and instrumentation. Out of the total 156 papers submitted, 122 papers were sent for double-blind review after preliminary scrutinization and plagiarism check. Out of these, 43 papers have been accepted for publication, and 13 of such papers are being published in this book as chapters.

We believe that the proceedings of the conference will be well received by researchers working in the domain and will be an inspiration for budding researchers to explore more into the varied domains in which the papers are presented. The papers presented in this *Lecture Notes in Electrical Engineering-Control*

Instrumentation System Conference (LNEE-CISCON) proceedings are mainly in the domains of process control, automation, instrumentation, robotics, image processing, and many more. The readers of this proceedings will get an insight into the varied areas in which contemporary research is being carried forward in this domain and get started to go ahead. These papers will give openings for the beginners and also the direction for those who are working in these specific domains already. We are confident that the proceedings will be accepted by prospective researchers very well and give encouragement for us to go ahead with organizing CISCON every year with lot many new ideas and scope.

We wish to take this opportunity to acknowledge the Council of Scientific and Industrial Research (CSIR), New Delhi, for financially supporting this event.

This event was made possible by the utmost support from Chancellor of MAHE Padmashree Awardee Dr. Ramadas M Pai, Pro-Chancellor Dr. H. S. Ballal, Vice Chancellor Dr. Vinod Bhat, Registrar Dr. Narayana Sabhahit, Chief Warden, Section Heads of finance, transport, accommodation, and other logistic services who deserve our heartfelt gratitude. The Director of Manipal Institute of Technology Dr. Srikanth Rao, Joint Director Dr. B. H. V. Pai, and the Head of the Department of Instrumentation and Control Engineering Dr. Dayananda Nayak deserve lots of appreciation for their constant guidance and motivation.

The conveners of the conference, Mr. Mukund Kumar Menon and Mr. P. Chenchu Saibabu, deserve a special recognition for their several months of untiring work toward this conference.

Special thanks to Dr. Santhosh K. V., Department of Instrumentation and Control Engineering, Manipal Institute of Technology, for coordinating with Springer Nature and enabling the proceedings to be published as Lecture Notes in Electrical Engineering—Proceedings of Control Instrumentation System Conference.

We express our sincere gratitude to the administrating staff of Manipal Academy of Higher Education (MAHE), Manipal Institute of Technology, and also the Department of Instrumentation and Control Engineering for their wholehearted support in making the conference event.

We sincerely acknowledge the unanimous technical reviewers and all contributing authors for taking time and effort to send their research work and adhering to all review comments and formatting requirements. We also wish to place our gratitude to Springer Nature for accepting our request to publish the accepted/presented papers in CISCON 2018.

Finally, we acknowledge all who have directly or indirectly helped us in organizing this event successfully and bringing out this proceedings.

Manipal, India
October 2018

Prof. Ravindra D. Gudi
Prof. C. Shreesha

Contents

Editors and Contributors

About the Editors

C. Shreesha is a Professor in the Department of Instrumentation and Control Engineering in Manipal Institute of Technology, India. Before joining MIT Manipal, he has worked at NMAMIT, Niite and served as HOD of E&E Engineering and Controller of Examinations. He received his Ph.D. from Indian Institute of Technology (IIT) Bombay, India for his work on Control Relevant Identification, and his research interests include linear, non-linear and optimal control, and Image Processing. Professor Shreesha is a Fellow of Institution of Engineers, India as well as a life member of the Indian Society for Technical Education (ISTE) and the Indian Society of Lighting Engineers. He has published more than 60 research papers in national and international journals and conferences.

Ravindra D. Gudi is a Professor and Head of the Department of Chemical Engineering in IIT Bombay. After completing his B. Tech. from IIT Bombay, he went on to do his Ph.D. from the University of Alberta, Canada. His research areas include control relevant identification, nonlinear identification, scheduling and decision support, disturbance/fault accommodation and optimal control of fermentation processes. He is the recipient of the Herdillia Award for Excellence in Basic Research by the Indian Institute of Chemical engineers, the Annual Technical Excellence Award in 2007 and 2008 by Honeywell, and the Manudhane Applied Research Award by IIT Bombay, among others. Professor Gudi holds 7 patents and has published 6 book chapters, and has more than 125 scopus indexed research publications.

Contributors

Sharad Bhartiya Department of Chemical Engineering, Indian Institute of Technology Bombay, Mumbai, India

C. R. Byrareddy Department of Electronics and Communication Engineering, BIT, Bangalore, Karnataka, India

C. Shreesha Department of Instrumentation and Control Engineering, Manipal Institute of Technology, Manipal Academy of Higher Education, Udupi, Karnataka, India

Ravindra D. Gudi Department of Chemical Engineering, Indian Institute of Technology Bombay, Mumbai, India

K. R. Guruprasad Department of Mechanical Engineering, National Institute of Technology Surathkal, Surathkal, Karnataka, India

T. N. Kesar ECE Department, DSCE, Bangalore, Karnataka, India

S. H. Laskar National Institute of Technology Silchar, Cachar, Assam, India; Sant Longowal Institute of Engineering and Technology, Longowal, Sangrur, Punjab, India

Kshetrimayum Lochan Department of Mechatronics, Manipal Institute of Technology, Manipal, Karnataka, India

Barasha Mali National Institute of Technology Silchar, Cachar, Assam, India; Sant Longowal Institute of Engineering and Technology, Longowal, Sangrur, Punjab, India

T. C. Manjunath Electronics Communication Engineering Department, Dayananda Sagara College of Engineering, Bangalore, Karnataka, India

Niyan Marchon Padre Conceicao College of Engineering, Verna, Goa, India

Doddabasappa Marebal Department of Automation and Robotics, KLE Technological University, Hubballi, India

P. M. Mohammad Minhaz Falaki Cellprop Pvt Ltd., Bengaluru, India

Gourish Naik Electronics Department, Goa University, Taleigao, Goa, India

Vishnu G. Nair Department of Aeronautical and Automobile Engineering, Manipal Institute of Technology, Manipal Academy of Higher Education, Manipal, Karnataka, India

Akshar Padman Department of Mechanical Engineering, National Institute of Technology Surathkal, Surathkal, Karnataka, India

Nandini A. Pattanshetti Department of Chemistry, Karnataka University Dharwad, Dharwad, Karnataka, India

Rajarshi Paul Intel Technologies India Pvt. Ltd., Bengaluru, India

G. Pavithra VTU Regional Resource Centre, Belagavi, Karnataka, India

P. S. Pravin Department of Chemical Engineering, Indian Institute of Technology Bombay, Mumbai, India

Binoy Krishna Roy Department of Electrical Engineering, National Institute of Technology Silchar, Silchar, Assam, India

Basavaraj S. Sannakashappanavar Department of Electronics and Telecommunication Engineering, ADCET, Ashta, Maharashtra, India

M. Shivakumar Department of EIE, GSSSIETW, Mysuru, India

Jay Prakash Singh Department of Electrical Engineering, Rewa Engineering College, Rewa, MP, India

M. V. Sreenivas Rao Department of EIE, GSSSIETW, Mysuru, India

Sanjit Varma Department of Electronics and Communication Engineering, SVEC, Tirupati, Andra Pradesh, India

Aniruddh Bahadur Yadav Department of Electronics and Communication Engineering, SVEC, Tirupati, Andra Pradesh, India

Dynamic Analysis of an Integrated Reformer-Membrane-Fuel Cell System with a Battery Backup and Switching Controller for Automotive Applications

P. S. Pravin, Ravindra D. Gudi and Sharad Bhartiya

Abstract Fuel cells have been considered as an ideal source of energy in the future power generation applications due to its pollution-free nature, noise-free operation and better efficiency. Direct storage of hydrogen in specially designed tanks for automobiles running on fuel cells is not a viable option due to several drawbacks associated with safety and space limitations. To overcome the challenges of direct onboard storage of hydrogen, storing hydrocarbons rich in hydrogen and suitably reforming it to produce hydrogen using several reforming techniques seems to be an acceptable option. Using available gas purification techniques such as palladium membrane-based gas separation, pure hydrogen gas can be extracted from a mixture of other gases and can be fed to the fuel cell for generating power. In this work, a mathematical model of the battery system is analyzed along with a switching controller operating based on an energy management policy. The switching controller switches between battery and fuel cell to ensure a delay-free delivery of the power to the external load. A case study on the dynamic behavior of the integrated system under set point changes in the power demand is analyzed in the presence of a battery backup and a switching controller.

Keywords Auto thermal reformer · Palladium membrane hydrogen separation · Polymer Electrolyte Membrane Fuel Cell (PEMFC) · Multi-loop control · Battery · Switching controller

---

P. S. Pravin · R. D. Gudi (✉) · S. Bhartiya (✉)
Department of Chemical Engineering, Indian Institute of Technology Bombay, Mumbai 400076, India
e-mail: pravin_ps@iitb.ac.in

R. D. Gudi
e-mail: ravigudi@iitb.ac.in

S. Bhartiya
e-mail: bhartiya@che.iitb.ac.in

C. Shreesha and R. D. Gudi (eds.), *Control Instrumentation Systems*, Lecture Notes in Electrical Engineering 581,
https://doi.org/10.1007/978-981-13-9419-5_1

1 Introduction

The primary objective of any power generating system is to generate power with lesser percentage of harmful emissions. Research is in progress to find a suitable alternative to internal combustion engines for automotive applications considering efficiency and manufacturing cost into account. A critical characteristic of any power generating system is to deliver power to the output load without significant delay. Performance in the transient condition is the most crucial factor considered while a power generation system is designed especially in automotive applications. Advances in fuel cell and battery research has been a breakthrough in the area of energy conversion and storage respectively in portable, stationary and automotive power applications. Majority of the currently available fuel cells utilize hydrogen as its fuel for power generation. Due to lack of efficient hydrogen distribution centers and limitations on direct storage of hydrogen onboard fuel cell vehicles, fuel cell technology has not been an attractive option in the automotive industry in many parts of the world. Cryogenic temperatures are crucial for storing liquid hydrogen as the hydrogen boiling point is −252.8 °C at a pressure of one atmosphere. Large composite vessels that can withstand high pressures are required for storage of hydrogen onboard in fuel cell vehicles [1]. A possible alternative for the direct hydrogen storage is to employ an onboard fuel processing and fuel purification subsystem that utilizes directly stored hydrocarbons rich in hydrogen along with the fuel cell. Due to unexpected traffic conditions and sudden acceleration/deceleration in an automobile, design should be done in order to avoid the expected significant lag between the demanded power by the output electric load and power generated by the fuel cell. It is to be noted that the response of fuel processing subsystem is sluggish compared to the frequency of power demand variations which results in slower dynamics of the overall system. This demands the necessity of an auxiliary power source like battery or super capacitor in order to ensure a delay-free response of the overall power generation system.

An auto thermal reformer is adopted as the fuel processing subsystem because an auto thermal reformer as compared to a steam reformer does not demand the need to supply or remove heat continuously to or from the system except during start of the reaction. Along with H_2, the by-product stream at the exit of an auto thermal reformer consists of other gases such as CO, CO_2, O_2, unconverted CH_4, H_2O (steam) and the inert N_2. Pure H_2 needs to be separated from the mixture of these gases by some mechanism. Among the various separation mechanisms available for gas separation, dense palladium membrane has been considered for H_2 gas purification in the present paper. High hydrogen permeability, fast hydrogen absorption and transport kinetics as well as excellent thermal stability of the palladium membrane qualifies it to be used as the gas purification sub system for fuel cell applications [2, 3]. Polymer electrolyte membrane fuel cell (PEMFC) is preferred as the power generation subsystem for the current work. Modeling and control studies on the integrated reformer membrane fuel cell system without a battery backup has already been reported in an earlier work [4]. Authors in [5] designed a power path controller utilizing ideal diodes that controls the flow of electricity in case of battery and fuel cell hybrid systems. The design of a hybrid power distribution source with a charge management system has been reported in the literature [6]. Authors in [7] proposed a fuel cell—battery—super capacitor hybrid power source to be used in fuel cell hybrid electric vehicles. In the present paper, we examine the overall dynamic behavior of the integrated reformer membrane fuel cell system with a battery backup and switching controller to explore the dynamic response of the system under a realistic scenario. Appropriate control structures are designed based on three single loop controllers and is evaluated for tracking the set point and rejecting the disturbances.

The paper is structured as follows. The next section presents a brief overview of the integrated system. The fuel processing, fuel purification, and power generation subsystems are discussed briefly in the following sections. A battery backup system is analyzed with the help of an equivalent circuit model. A switching controller that selects between battery and fuel cell operating on an energy management policy is presented in the next section. A case study on the dynamics associated with the integrated system with fluctuating load power demand and the performance of the designed controller tracking the set point trajectory is presented followed by results and discussions with conclusion as the last section.

2 System Description

The individual subsystems are combined together to form a single integrated structure in order to avoid the drawbacks of direct onboard storage of hydrogen in fuel cell portable applications [4]. Figure 1 shows a block diagram of the integrated system with necessary ancillary units along with a battery backup and a switching controller.

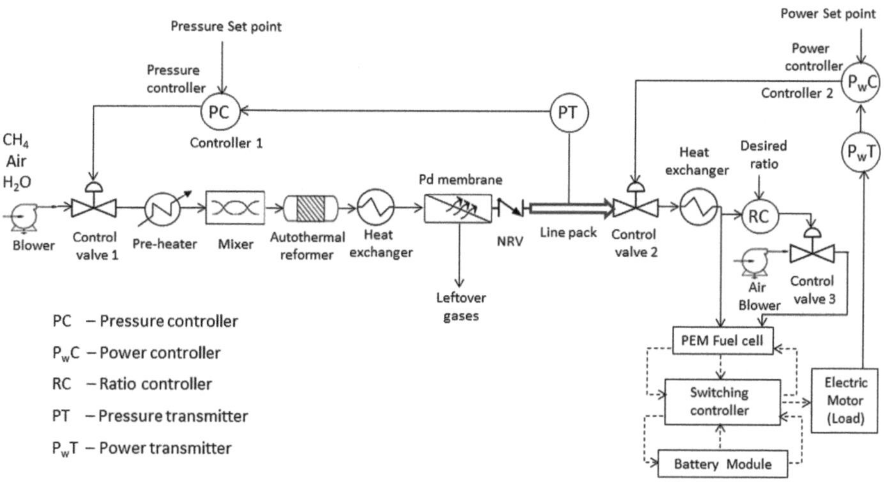

Fig. 1 Block diagram of the integrated system with battery backup

Requisite amount of stored methane, steam and air are preheated to appropriate reforming temperatures and properly mixed to be delivered as inputs to the reformer. The output of auto thermal reformer contains hydrogen along with mixture of other gases. A palladium membrane separation unit separates hydrogen from other gas mixtures so as to feed pure hydrogen gas through a non-return valve (NRV) and control valve 2 to anode inlet of the PEM fuel cell to generate power. Heat exchangers are installed at appropriate locations to sustain the intermediate temperatures of gas at desirable values. It is assumed that both the heat exchangers are under perfect control. As the exothermic reactions are dominated compared to the endothermic reactions in an auto thermal reactor, the need of a pre-heater is mandatory during start of the reaction. During the plant start-up, the gas pipeline coupling the NRV and control valve 2 contains appropriate amount of hydrogen so that the line pack pressure is maintained at 1 atm.

Three controllers, one for regulating flow of hydrogen at the anode inlet of the fuel cell, second for regulating flow of methane at the inlet side of reformer and third for regulating air to hydrogen ratio in PEMFC are designed to effectively control the integrated system. Based on the demanded power at the output load, controller 2 (Proportional controller with $K_P = 0.1$) manipulates the hydrogen flow rate to the anode inlet of the fuel cell using control valve 2. Due to control valve 2 opening, pressure in the line pack gets disrupted. This activates the controller 1 (PID controller with $K_P = 0.3$, $K_I = 0.0001$ and $K_D = 0.4$) to manipulate the methane inlet flow rate using control valve 1 so as to bring the line pack hydrogen pressure (or molar concentration of hydrogen in the line pack) back to the set point value. A switching controller toggles between battery and fuel cell to deliver power to the vehicle and charge/discharge the battery based on the output load demand, quantity of hydrogen gas in the line pack and the battery charge status.

3 Fuel Processing Subsystem: Auto Thermal Reformer

As already discussed, an auto thermal reformer is preferred as the fuel processing subsystem, wherein appropriate quantity of methane, air and steam are fed as input to generate hydrogen along with mixture of other gases. The molar ratio of steam to carbon is chosen to be 1:1 and that of oxygen to carbon to be 0.45:1. Among many reactions possible to exist in an auto thermal reformer, only the dominant ones are considered in the present study as can be seen from Table 1 [4]. A reactor with a cylindrical shape with a length of 0.2 m is considered. Nickel is chosen as the catalyst with a density value of 1870 kg/m^3. For simulation studies, a one-dimensional dynamic model of an auto thermal reformer is adopted from [8]. For getting an idea of the expressions for mass and energy balance, reaction rates of each species and values of the constants used, the reader is directed to refer [8].

Table 1 Auto thermal reforming reactions [8]

	Reactions	$\Delta H_{298\,K}$ (kJ/mol)
Steam Reforming	$CH_4 + H_2O \leftrightarrow CO + 3H_2$	206.2
Steam Reforming	$CH_4 + 2H_2O \leftrightarrow CO_2 + 4H_2$	164.9
Water gas shift	$CO + H_2O \leftrightarrow CO_2 + H_2$	−41.1
Total Combustion	$CH_4 + 2O_2 \leftrightarrow CO_2 + 2H_2O$	−802.7

4 Fuel Purification Subsystem: Palladium-Based Membrane Separation

At the anode inlet, availability of pure hydrogen gas is a mandatory requirement for efficient operation of a PEMFC. The efficiency and lifetime of the fuel cell can adversely get affected by impurities present in the input feed gas. Regardless of whether the application is for stationary or portable, palladium membrane-based hydrogen gas purification seems to be an excellent option [9]. In this simulation study, it is assumed that the membrane is free of any defects without any physical porous support. A term θ_{mem} is included in the permeation equation which denotes the poisoning of membrane caused due to CO adsorption on to the membrane surface. Thus, $(1 - \theta_{mem})$ indicates the membrane fraction literally available for separation of hydrogen gas. This can partially obstruct or prevent the flow of hydrogen gas through the palladium membrane. In this study, it is assumed that the permeability of membrane for all other gases are zero except for hydrogen.

5 Power Generation Subsystem: PEMFC and Battery

5.1 PEMFC

Hydrogen gas after further purification by the palladium membrane is fed as input to the PEMFC anode inlet to generate power. Air after necessary humidification is given to the cathode inlet of the fuel cell. As soon as the active catalyst sites comes in contact with the reactants, electrochemical reactions as shown in Table 2 initiate. Nafion is used as the membrane material which require proper hydration for better conductivity

Table 2 Electrochemical reactions in PEMFC

Electrode	Reactions
Anode	$H_2 \leftrightarrow 2H^+ + 2e^-$
Cathode	$\frac{1}{2}O_2 + 2H^+ + 2e^- \leftrightarrow H_2O$
Overall	$H_2 + \frac{1}{2}O_2 \leftrightarrow H_2O$

of the H^+ ions through it. This requirement of hydration of the membrane restricts the operation of PEMFC's at temperatures below 90 °C. An open circuit voltage of about 1 V is usually achieved by a single fuel cell and under load, it can drop down to 0.6–0.7 V. However, this small voltage does not satisfy the voltage requirement for real-world applications that actually require several hundreds of volts. For achieving these high voltages, individual cells are arranged in series to form a fuel cell stack so that the aggregate voltage matches the required voltage value. More information on the mathematical model equations of a low temperature PEMFC can be obtained from Ref. [10].

5.2 Battery

In fuel cell powered automobiles or other power applications, there is always a time lag associated with the power delivered by the fuel cell and power demanded by the electric load. This time delay is unacceptable especially for automotive applications that involves fast transients and sudden load demand variations. To ensure satisfactory performance of the designed system, supplementary power sources such as batteries and super capacitors having much rapid response time compared to fuel cell is essential. The major parameter to be considered while designing a battery system for power applications is the State-of-charge (SOC). It is defined as the ratio of the remaining capacity to the fully charged capacity [11]. Change in SOC of a battery for a time interval ΔT can be expressed as follows.

$$SOC(t + \Delta T) = SOC(t) - \frac{\text{Used capacity}}{\text{Total capacity}} \tag{1}$$

$$\text{Used capacity} = \int I.\Delta T \tag{2}$$

Total capacity is the rated Ah capacity of the battery. For the present study, a battery with a capacity rating of 22 Ah is considered [11]. Ideally, it can be viewed that SOC of a battery will be one when the battery is charged fully and zero when it is discharged to a critical voltage. It is a usual practice to maintain the SOC at a value between 0.5 and 0.7 [12]. Authors in [11] modeled battery as a pure resistance source in which they used battery open circuit voltage V_{oc} and internal resistance R, for each specific SOC to calculate the battery terminal voltage $V_{terminal}$ and output current I. The authors specify that the values of V_{oc} and R is not generally available from the battery specifications and are also difficult to measure in real time. An empirical relation between various parameters like $V_{terminal}$ and I corresponding to a given SOC was developed by the authors in [11]. In this work, we have adopted this empirical relationship for the battery model as follows.

$$V_{terminal} = a * I + b + c * SOC \tag{3}$$

Specifically,
a = −0.18, b = 172, c = 40 for $SOC < 0.4$
a = −0.20, b = 186.94, c = 4 for $1 > SOC >= 0.4$.

Expression for power demanded from the battery is given by

$$P_{dem} = I * V_{terminal} \tag{4}$$

Substituting the expression for $V_{terminal}$ from Eq. (3) in Eq. (4), we obtain the expression to calculate current I given by

$$I = \frac{-(b + c * SOC) + \sqrt{(b + c * SOC)^2 + 4 * a * P_{dem}}}{2 * a} \tag{5}$$

For a given SOC of the battery and power demanded by the electric load from the battery, battery charge/discharge current I can be calculated. I is assumed to be positive while discharging and negative while charging.

6 Switching Controller

The switching controller switches among the battery and fuel cell in order to realize a delay-free response of the integrated system. Main reason for the time lag between the power demanded by the output load and the power delivered by the fuel cell system is due to the sluggish behavior of the highly complex reformer system. Another reason can be due to the dynamics associated with the control valve 2. The logic component of the switching controller is realized using if-then-else rules based on the energy management policy given in Table 3. The switching controller manages the toggling between battery and fuel cell to ensure a delay-free response. P_1 denotes the line pack hydrogen pressure, P_2 denotes the fuel cell pressure (1 atm), SOC_{bat} indicates the State of Charge of the battery at a given time instant, SOC_{low} and SOC_{high} denotes the SOC lower and upper limits of the battery respectively. During the plant start-up, SOC_{low} and SOC_{high} values are set equal to 0.2 and 0.8 respectively. P_{FC} and P_D denotes the power delivered by the fuel cell and power demanded by the electric load in kW respectively.

Table 3 Energy management policy of switching controller

Logic	Condition	Status
L_1	$P_1 <= 1.3 * P_2$ & $SOC_{bat} < SOC_{low}$	Battery OFF Fuel cell OFF
L_2	$P_1 <= 1.3 * P_2$ & $SOC_{bat} > SOC_{low}$ & $abs(P_{FC} - P_D) > 0.015$	Battery ON Fuel cell OFF
L_3	$P_1 > 1.3 * P_2$ & $SOC_{bat} < SOC_{high}$ & $abs(P_{FC} - P_D) < 0.015$	Battery CHARGING Fuel cell ON
L_4	$P_1 > 1.3 * P_2$ & $SOC_{bat} < SOC_{low}$ & $abs(P_{FC} - P_D) > 0.015$	Battery IDLE Fuel cell ON
L_5	$P_1 > 1.3 * P_2$ & $SOC_{bat} > SOC_{low}$ & $abs(P_{FC} - P_D) > 0.015$	Battery ON Fuel cell ON
L_6	$P_1 > 1.3 * P_2$ & $SOC_{bat} >= SOC_{high}$ & $abs(P_{FC} - P_D) < 0.015$	Battery CHARGE FULL Fuel cell ON

7 Case Study: Dynamic Analysis of the Integrated System with Battery Backup and Switching Controller

This sections discusses the dynamic behavior of the integrated system with respect to set point changes in the demanded power by the fuel cell vehicle. The percentage opening of control valve 1 determines the rate of hydrogen production at the reformer side. Opening of control valve 2 is permitted only when the upstream line pack pressure (P_1) attains a value 1.3 times more than the downstream fuel cell pressure (P_2). Percentage opening of control valve 3 determines the air to hydrogen ratio to be fed as inputs to the anode and cathode of the fuel cell. It is assumed that during start of the simulation, the plant is at steady state condition with all the variables at its steady state value. As already discussed, a multi loop control strategy is designed and implemented wherein fuel cell load power demand and hydrogen molar concentration in the line pack (or line pack hydrogen pressure assuming ideal gas law) are chosen to be the controlled variables. Condition L_1 becomes active when there is no minimum charge stored in the battery as well as no sufficient hydrogen existing in the line pack. Logic L_2 gets active when there is sufficient battery charge while there is no adequate fuel in the line pack. In this case, battery continues to supply power to the external load with fuel cell in OFF state. When the line pack gets filled with sufficient hydrogen to achieve a pressure (P_1) greater than 1.3 times the downstream pressure (P_2), condition L_3 gets activated wherein fuel cell starts delivering power to the load as well as starts charging the battery. This will continue until the difference between the power delivered by the fuel cell (P_{FC}) and power requested by the electric load (P_D) becomes less than an acceptable threshold value of 0.015 kW. Whenever SOC_{bat} becomes less than SOC_{low}, battery is no longer in a state to deliver power and the condition L_4 gets activated. When the condition is favorable for both the battery and fuel cell to be ON and if the difference between P_{FC} and P_D is greater than an acceptable tolerance value, condition L_5 gets activated. It

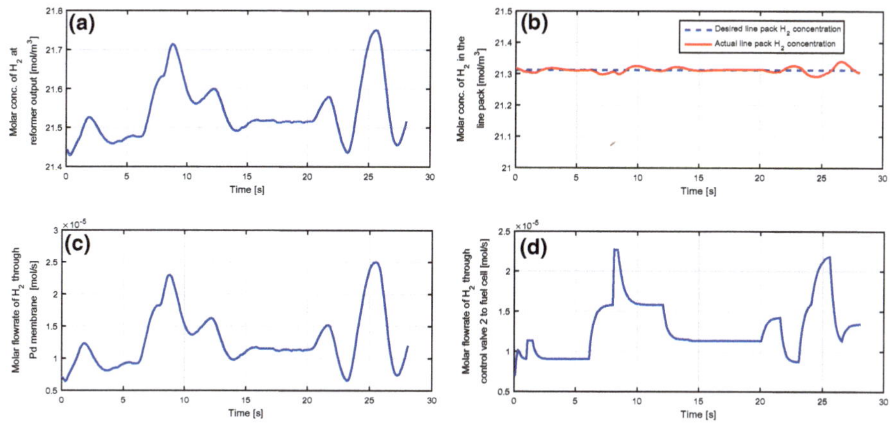

Fig. 2 **a** Molar concentration of H_2 at exit of reformer **b** performance of controller 1 maintaining a constant H_2 molar concentration in the line pack **c** molar flow rate of H_2 through Pd membrane **d** molar flow rate of H_2 to fuel cell through control valve 2

is considered that at steady state condition of the plant, operating policy L_6 is active illustrating that the battery is fully charged and fuel cell alone is delivering power to the output load.

The dynamic profile of the molar concentration of hydrogen at the reformer exit is as shown in Fig. 2a. It can be seen that at $t = 0$, the molar concentration starts from its steady state value of 21.4 mol/m^3 and starts varying depending on the set point changes in the power output. The main aim of controller 1 is to maintain a constant H_2 molar concentration in the line pack irrespective of the set point changes in the output load power. Fig. 2b shows the profiles for the desired and actual values of molar concentration of H_2 in the line pack illustrating the effectiveness of the designed controller. It can be seen that at start of the simulation, the actual H_2 molar concentration in the line pack was exactly equal to the desired value. In response to the set point changes in the output power, the actual line pack concentration deviates from the desired value. Controller 1 detects this deviation to regulate the methane flow rate using control valve 1 so as to bring the actual molar concentration back to the desired value. The profiles showing the molar flow rate of H_2 through the palladium membrane and through control valve 2 to fuel cell are given in Fig. 2c and Fig. 2d respectively. The percentage opening of control valve 1 and control valve 2 is shown in Fig. 3a and Fig. 3b depending on the command from controller 1 and controller 2 respectively. At start of the simulation, logic L_6 was active with battery fully charged and fuel cell alone supplying power to the output load. Based on the energy management policy discussed in Table 3, the switching controller switches between battery and fuel cell to provide a delay free response to the output load. As can be seen from Fig. 3c, based on the set point power requested by the electric load, switching controller connects/disconnects the battery in the loop. It can be interpreted that at any given point of time, the sum of power delivered by the battery

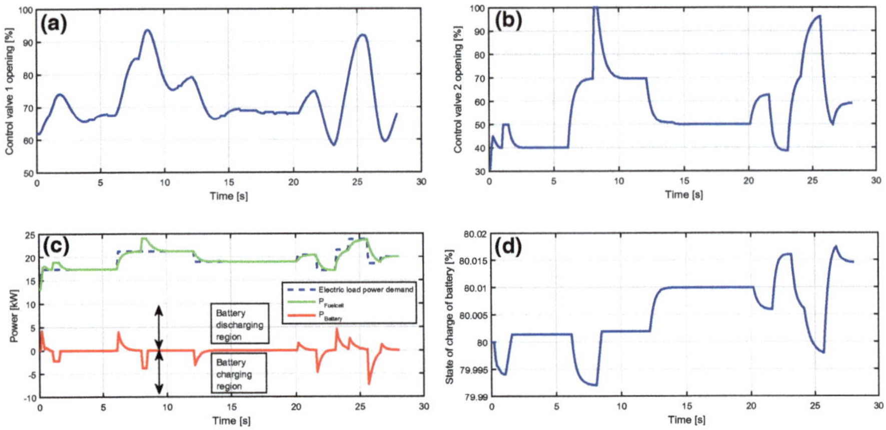

Fig. 3 **a** Percentage opening of control valve 1 **b** percentage opening of control valve 2 **c** power profile showing performance of switching controller and multi loop controllers **d** state of charge of the battery

and that given by the fuel cell is exactly equal to the set point power demanded by the external load. It is to be noted that, whenever there is a sudden decrease in the power demand, the additional power generated by the fuel cell is utilized to charge the battery as can be seen from Fig. 3c. In a scenario where the battery was already fully charged (ie. $SOC_{bat} = SOC_{high}$), then under sudden decrease in the power demand, the SOC_{high} value is raised from 0.8 to 0.9 so that the additional power generated by the fuel cell can be utilized to charge the battery beyond the SOC_{high} value. The change in SOC value depending on the varying switching action of the switching controller can be seen from Fig. 3d. As obvious, it can be noticed that the SOC decreases during battery discharging and increases during battery charging.

8 Conclusions

The dynamics of a reformer unit is at least an order of magnitude slower compared to the fuel cell system. This work examines the feasibility of combining reformer with a palladium membrane separation unit and a fuel cell unit with a battery backup to explore its dynamic behavior under realistic scenarios. A multi loop control strategy was implemented with conventional PID controllers for which power demanded by the electric load and molar concentration of hydrogen in the line pack were chosen as the controlled variables. A mathematical model of the battery is adopted from the literature which gives the relation between the state of charge and power demanded from the battery. A switching controller is designed that operates based on some energy management policy that toggles between battery and fuel cell. The integrated system is simulated with changes in the set point power demand and the simulation results show a delay free delivery of power to the electric load with negligible offset substantiating the efficacy of the designed fuel cell battery system.

References

1. Qi A, Peppley B, Karan K (2007) Integrated fuel processors for fuel cell application: a review. Fuel Process Technol 88(1):3–22
2. Iwuchukwu IJ, Sheth A (2008) Mathematical modeling of high temperature and high-pressure dense membrane separation of hydrogen from gasification. Chem Eng Process Process Intensif 47(8):1292–1304
3. Okazaki J, Ikeda T, Tanaka DAP, Sato K, Suzuki TM, Mizukami F (2011) An investigation of thermal stability of thin palladium silver alloy membranes for high temperature hydrogen separation. J Membr Sci 366(1):212–219
4. Pravin PS, Gudi RD, Bhartiya S (2018) Dynamic modeling and control of an integrated reformer-membrane-fuel cell system. Processes 6(9)
5. Howroyd S, Chen R (2016) Powerpath controller for fuel cell & battery hybridisation. Int J Hydrog Energy 41(7):4229–4238
6. Blackwelder MJ, Dougal RA (2004) Power coordination in a fuel cell—battery hybrid power source using commercial power controller circuits. J Power Sources 134:139–147
7. Fathabadi H (2018) Novel fuel cell/battery/supercapacitor hybrid power source for fuel cell hybrid electric vehicles. Energy 143:467–477
8. Halabi M, de Croon M, van der Schaaf J, Cobden P, Schouten J (2008) Modeling and analysis of autothermal reforming of methane to hydrogen in a fixed bed reformer. Chem Eng J 137(3):568–578
9. Pinacci P, Drago F (2012) Influence of the support on permeation of palladium composite membranes in presence of sweep gas. Catalysis Today 193 (1):186–193. Proceedings of the 10th international conference on catalysis in membrane reactors
10. Methekar R, Prasad V, Gudi R (2007) Dynamic analysis and linear control strategies for proton exchange membrane fuel cell using a distributed parameter model. J Power Sources 165:152–170
11. He X, Hodgson JW (2002) Modeling and simulation for hybrid electric vehicles. I. Modeling. IEEE Trans Intell Transp Syst 3(4):235–243
12. Kolavennu P, Telotte J, Palanki S (2009) Analysis of battery backup and switching controller for a fuel-cell powered automobile. Int J Hydrog Energy 34:380–387

Design and Implementation of Fuzzy Logic Controller on MPSoC FPGA for Shell and Tube Heat Exchanger

Rajarshi Paul and C. Shreesha

Abstract In this paper design of fuzzy logic controller (FLC) on system on chip field programmable gate array (SoC FPGA) for Shell and Tube Heat Exchanger (STHE) is described. STHE is interfaced with Compact Reconfigurable Input Output (CRIO) based FPGA through data acquisition (DAQ) card. For designing controller on chip, National Instruments (NI) CRIO-9101 is reconfigured with FLC code. Interfacing is carried out with NI analog input-output device embedded with NI CRIO-9012 microcontroller and NI CRIO-9101 FPGA module. A graphical program is developed for real-time control of the process plant using CRIO based FPGA. FPGA acts as a standalone processor instead of PC for controlling the STHE. The fuzzy logic controller designed on FPGA is compared with benchmark controller in real time temperature control of STHE. It is demonstrated that the controller proposed outperformed conventional controller.

Keywords Fuzzy logic control · Embedded system · STHE · LabVIEW · FPGA · MPSoC

1 Introduction

Multi-Processor System-on-Chip (MPSoC) is one of the recent trends towards embedded system design. SoC field programmable gate array (FPGA) provides many advantages like higher integration and easy upgrades of systems in the field. Custom instructions and function units are enabled through a soft CPU core. Enhanced SoC development, debugging, testing and tuning are possible through reconfiguration. Intelligent design of control is a generic term for employing soft-computation

R. Paul
Intel Technologies India Pvt. Ltd., Bengaluru, India

C. Shreesha (✉)
Department of Instrumentation and Control Engineering, Manipal Institute of Technology, Manipal Academy of Higher Education, Udupi 576104, Karnataka, India
e-mail: shreesha.c@manipal.edu

C. Shreesha and R. D. Gudi (eds.), *Control Instrumentation Systems*,
Lecture Notes in Electrical Engineering 581,
https://doi.org/10.1007/978-981-13-9419-5_2

methodology. The work presented explores more on specific intelligent controller design on MPSoC FPGA.

Several researches are carried out on the FPGA implementation on various automotive plants. Only a few works are reported on embedded controller design for process plants. Huang et al. proposed a FPGA implementation for autonomous omnidirectional robotic movement [1]. For FPGA hardware programming they used the intellectual property of software packages. Meoung et al. carried out the work for SoC based embedded controller design using Altera FPGA [2]. Design and satisfactory performance of controller on chip are challenging which demands proper scheduling of tasks on FPGA and synchronization with the system clock. Hassan et al. discussed about PID controller design on MPSoC and multicore microcontrollers [3]. FPGA's logical cells programmed for a task is dependent on the synthesizing of the code. Sun et al. discussed in their work software design flow along with FPGA implementation for the biomedical application [4].

Recent research explores more about the PID controller design on FPGA for various motor control or other control applications [5–8]. However very few publications could only be seen on Fuzzy based controller on FPGA for heat exchanger. The reason for working with fuzzy logic controller (FLC) is highlighted and it is chosen as the suitable controller for reconfiguring the FPGA's logical elements. Fuzzy logic control provides robust mechanism than the conventional method with system parameter variation. Work presented in this paper discusses MPSoC FPGA FLC design for STHE process.

FPGA may be programmed with any conventional hardware description language (HDL) [5, 9]. In the technique of coding with earlier HDL, program designer needs to spend more time and interfacing with the FPGA board was a challenging task. Alternatively LabVIEW provides easier methods for FPGA programming with the inbuilt software tools. For software development LabVIEW platform is extensively used. In the proposed work LabVIEW inbuilt integrated modules are used for interfacing with real-time plant appended with fuzzy logic.

For designing controller on chip, NI CRIO-9102 and NI CRIO 9101 real-time controller is reconfigured with intelligent controller code. The interfacing is carried out with NI analog input-output device embedded with NI CRIO-9012 microcontroller and NI-9101chassis. From the results of implementation, it is noticed that fuzzy controller performs better than conventional method of control [10].

Section 1 of the paper explains background of MPSoC FPGA. In Sect. 2 process of configuring the system on chip and parallel processing concepts adapted for the work are discussed. Section 3 explains about the CRIO programming logic for the proposed controller execution. Section 4 explores the implementation of fuzzy logic controller in real-time on SoC FPGA. Section 5 analyses results obtained and performance evaluation of proposed controller with the benchmark controllers.

2 MPSoC FPGA and Hardware Configuration

MPSoC is a comprehensive VLSI system which deploys various programmable processors on a single chip [13]. These processors work simultaneously for the destined application along with other system components. While SoC may also be designed with microcontroller, FPGA has the advantage that it can handle complex logic with simple soft core logic, has faster execution and less power consumption. CRIO 9101 is chosen for the FPGA module implementation by utilizing inbuilt ARM Cortex A9 processor of CRIO9012 module. In the work presented controller logic is kept within FPGA and continuous data monitoring is embedded with CRIO RT Processor.

LabVIEW FPGA compilation process is different from the traditional FPGA programming [11, 14]. With Xilinx compilation tool in LabVIEW embedded environment, the VHDL code was generated. The compiler synthesizes the VHDL code for place and route of components. After successful compilation, software generates the 'bit-file'. Bit-file will be downloaded to the FPGA and logical cells of hardware are configured.

2.1 *Configuring the MPSoC Based FPGA*

To meet objectives for controlling the STHE plant CRIO 9012 and CRIO 9101 are used simultaneously. Using LabVIEW Real-Time Module 8.6 software and with NI Scan engine, control system is developed. With floating point arithmetic capability real-time processor generates precise and definite output. FPGA controls the tasks using master clock. Conceptual view of the architecture of FPGA based multiprocessor system on chip is shown in Fig. 1. The human–machine interface design flow for RT control is represented in Fig. 2.

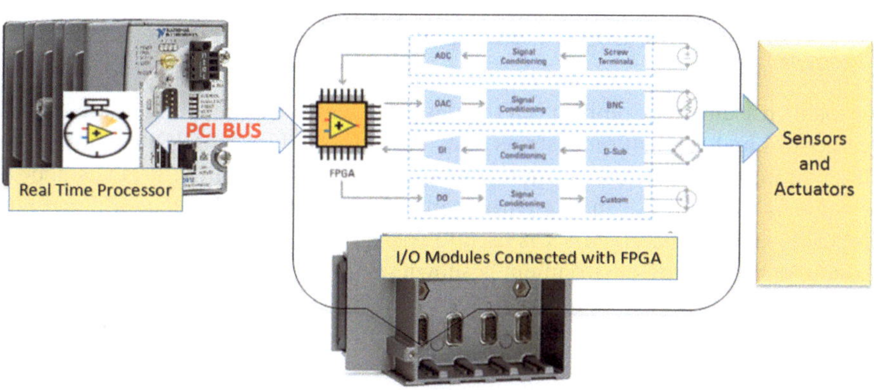

Fig. 1 CRIO 9012 and CRIO 9101 FPGA device together forms the MPSoC architecture

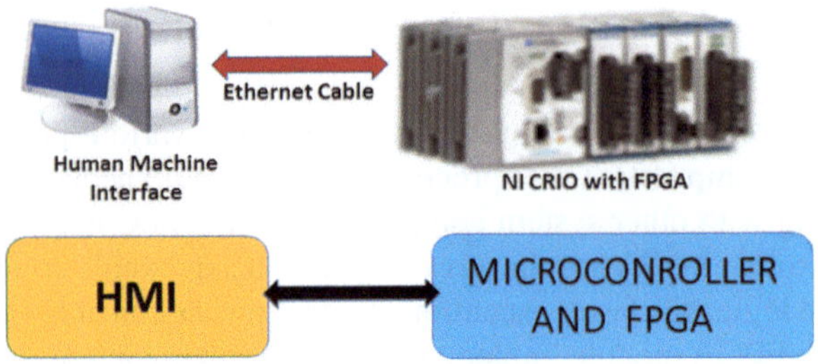

Fig. 2 Human machine interface design flow for RT control

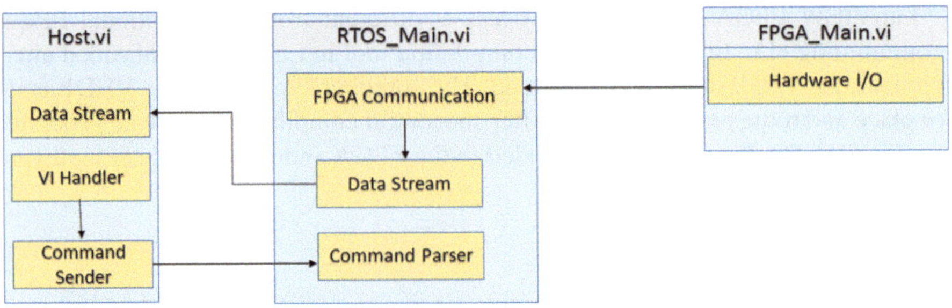

Fig. 3 Basic components of CRIO data communication for embedded monitoring

The crux of operation of this embedded system is dependent on FPGA. FPGA timing and synchronization for each task is maintained by a single clock in FPGA. With designed CRIO module's direct connection to FPGA, the latency for system response is null. The user interface also allows operator to configure hardware in order to set up communication with host and RTOS (Fig. 3).

3 Execution and Logic Description for Configuring CRIO

National Instruments Compact RIO is one of the advanced reconfigurable embedded control and acquisition systems which are powered by NI RIO technology for ultrahigh performance, user customization and re-configurability. It is designed to perform in harshest industrial environments. NI LabVIEW FPGA configured with proper IP settings and NI MAX is used to detect the CRIO9012 hardware. In Fig. 4 detailed programming environment is shown.

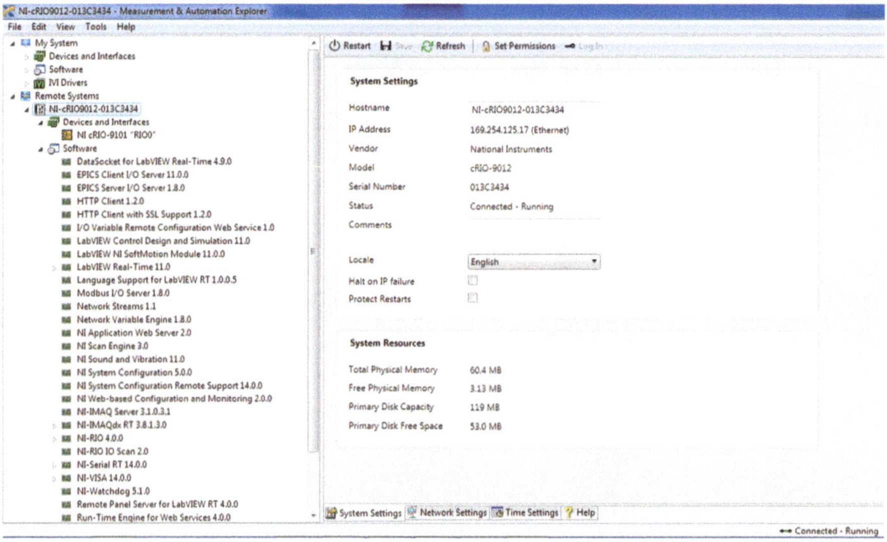

Fig. 4 NI Max hardware programming successfully recognizing the CRIO modules

3.1 Sending Command and Receiving Data from Real Time Host to FPGA

For advanced high-speed data transfer in a SoC, Direct Memory Access (DMA) technique is adopted [15]. It is an important entity for improving speed of communication which utilizes asynchronous First in First out (FIFO) technique. In the program developed\real-time FIFO configured for buffering data. After every shared variable is read the element is removed from FIFO. The high speed data are streamed between FPGA and realtime host using DMA. The FIFO created was meant for data flow from the buffer of FPGA target to real-time host. To read larger set of data or faster from the host buffer size is kept at higher value. RS232 port is accessed from the DAQ card of STHE process plant. Figure 5 shows the read FIFO and write FIFO configuration under FPGA design tree. The read and write FIFO stores 1029 elements in its block memory with slice fabric of reconfigurable chip. The programming environment of LabVEW11.0 for LVRT module accessing FPGA FIFO through DMA technique is shown in Fig. 6.

4 Program Logic for Data Acquisition and Control

FPGA main VI has character read/write which communicates with host VI using reference variable declared in other loops. In LabVIEW programming, logical representation is depicted in the block diagram. LabVIEW FPGA based CRIO is used to communicate with RS232. The protocol is setup for serial communication of STHE.

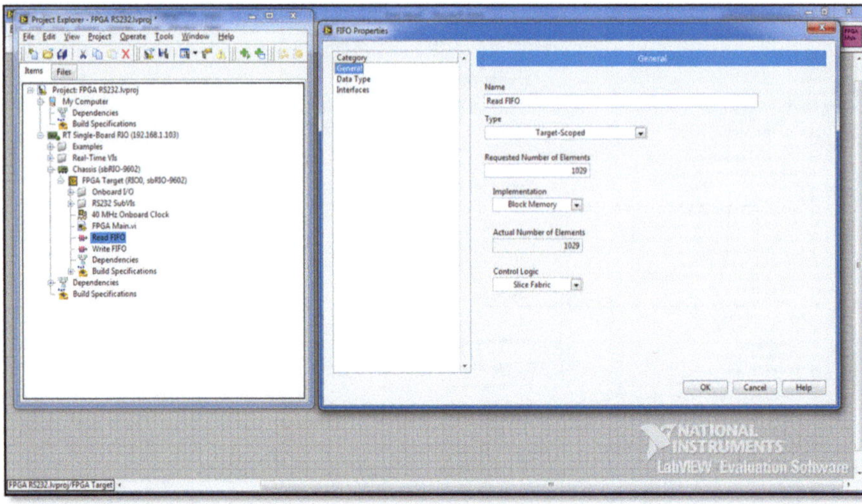

Fig. 5 FIFO has been implemented for FPGA memory access

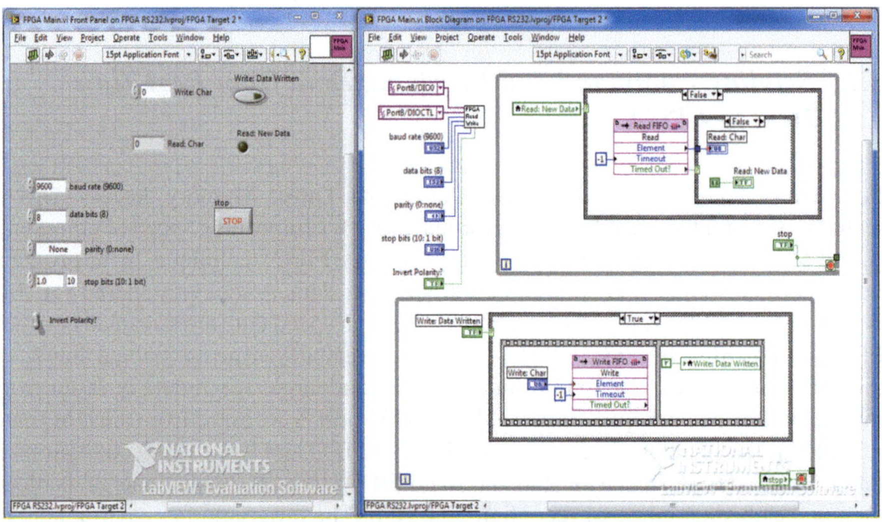

Fig. 6 LVRT module configuration created by invoking the FIFO modules of FPGA target

FPGA VIs runs on FPGA target. After creating an FPGA VI, this VI is compiled into an application before downloading and executing it on the target. LabVIEW uses inbuilt Xilinx Compilation tool to compile and download FPGA VI to FPGA target. The VI automatically creates a build specification and compiles them on FPGA target. Since the interactive front panel communication is set up for STHE, FPGA target runs VI on flash memory. To specify unique values for individual channels, initialization loop on the host VI is created.

4.1 Processing Loop on the Host VI

The processing loop on host VI was created to modify parameters or reset channels asynchronously during execution of FPGA VI. Software design flow for FPGA communication to the host and real-time target is shown in Fig. 7.

The code (VI) developed for serial communication was operated with the baud rate of 9600 with one stop bit with a time out of 1000 s and 8 data bits are transferred. The LabVIEW front panel for this subVI 'FPGA main ref out' variable does the subroutine call (see Fig. 8). The Project file in Fig. 9, contains two virtual folders, Read-Write and Real-Time VIs.

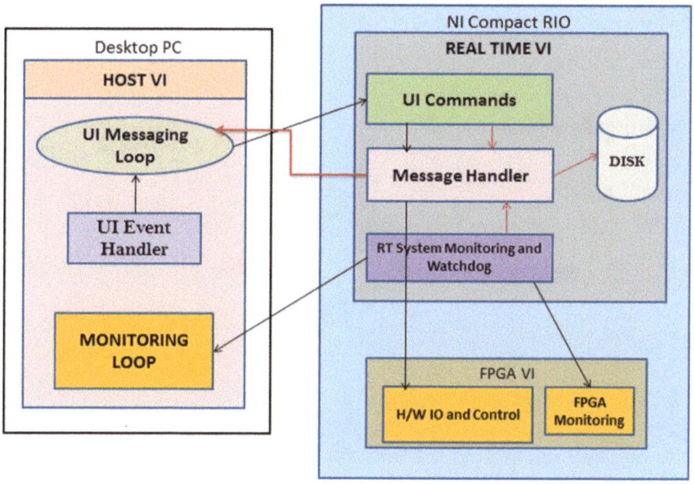

Fig. 7 LabVIEW RT target and Host communication with FPGA—work flow diagram

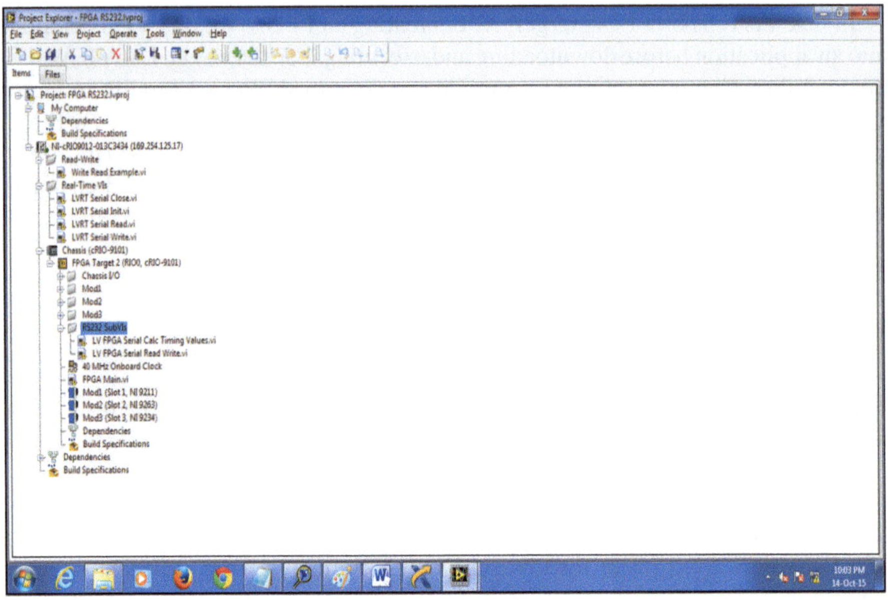

Fig. 8 The project window shows the FPGA VI and all hardware devices detected within FPGA

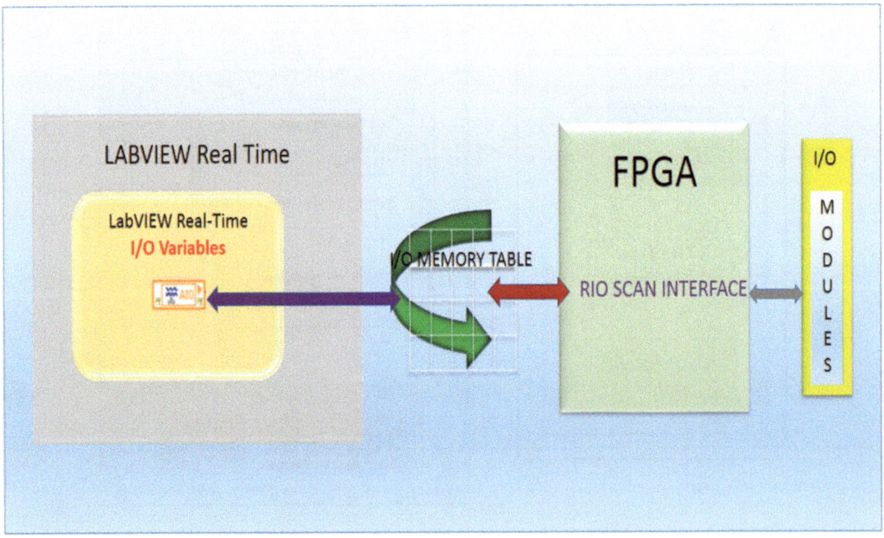

Fig. 9 Software flow diagram of CRIO scan interface

4.2 Implementing FLC on FPGA in Real Time

To run fuzzy controller through FPGA, CRIO9012 is configured first. Once the bit files successfully configure logical cells of FPGA, this acts as standalone processor [16–18]. The processor accesses I/O modules through scan interface mode of operation. I/O modules are automatically read and placed in the memory table using DMA FIFO mechanism. The whole process of data communication is shown in Fig. 9. The VI acquires signals from STHE and communicates with RS232 protocol. FPGA target interacts with RS232 VI and monitors the temperature control. On FPGA target two things occur in parallel-monitoring of the control loop and communicating with the Host VIs.

In this project for controller design three codes (VI) run on three different targets—the desktop PC, the Compact RIO Real-Time target and the Compact RIO FPGA target. On FPGA target, two tasks are running in parallel: executing the control loop and executing the watchdog loop. The successful code synthesis for FPGA is performed by inbuilt Xilinx compilation tool version 10.1 (see Fig. 10). The device utilization summary is shown in Table 1.

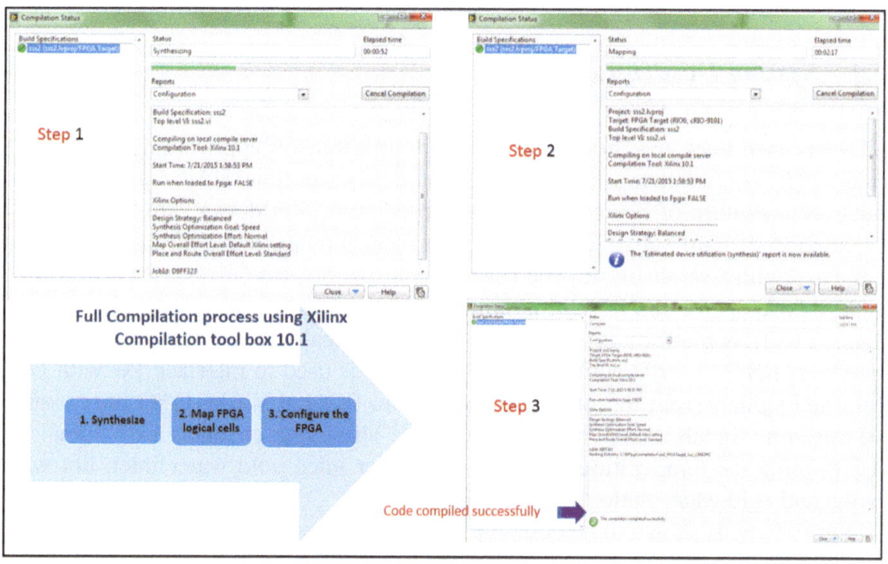

Fig. 10 FPGA code compilation process and bit file generation for configuring the hardware

Table 1 Xilinx FPGA device utilization summary

Logic utilization	Available	Used	Utilization (%)
No. slice registers	301270	1956	1
No. of sliced flip flops	4727	1258	27
No. bounded I/Os	500	354	70
No. of block RAM/FIFO	406	4	1

5 Fuzzy Logic Controller—Detailed Logic Description

In this section, LabVIEW based fuzzy logic controller design for real-time temperature monitoring of STHE will be discussed. Objective of implementation is to minimize error between actual outlet temperature and the setpoint to achieve efficient control of temperature. The fuzzy rule table forms associative memory by calculating error between set point and process variable. Controller output is noticed to be satisfactory with the regulation of hot water outlet temperature to desired values.

5.1 System Description

STHE is used in process control industry for heat transfer applications. In the work presented, manipulated variable is selected as cold water flow rate to maintain desired outlet temperature of hot water. The proposed controller doesn't involve tiresome calculation of gain values as in PID controller. Human linguistic expressions are emulated with a series of "IF-Then" rules.

The developed LabVIEW program is able to run hardware as well as to capture data. LabVIEW platform is used to produce a graphical view (see Fig. 11) of the user interface. RS-232 serial communication channel is used to interface PC with DAQ card and signals from the plant are acquired to PC. LabVIEW DAQ max express VI transmits signals acquired from 4 channels. These four channels are—signals for determining the temperature values of hot water inlet, cold water inlet, hot water outlet and cold water outlet from STHE.

Fuzzy logic is shown to be effective in real time control of STHE [12]. The fuzzy inferencing was done by fuzzification, rule evaluation and defuzzification. Fuzzy rules were developed using Mamdani's method. It is in "If Then" format. In Fig. 12 front panel of LabVIEW code showing fuzzy rules invoked during the process run is presented.

The fuzzy inference system associates each input to output using fuzzy logic. Mamdani's method is used for the formation of rules. Triangular membership function is utilized for the linguistic rules formation. In order to assign values to mem-

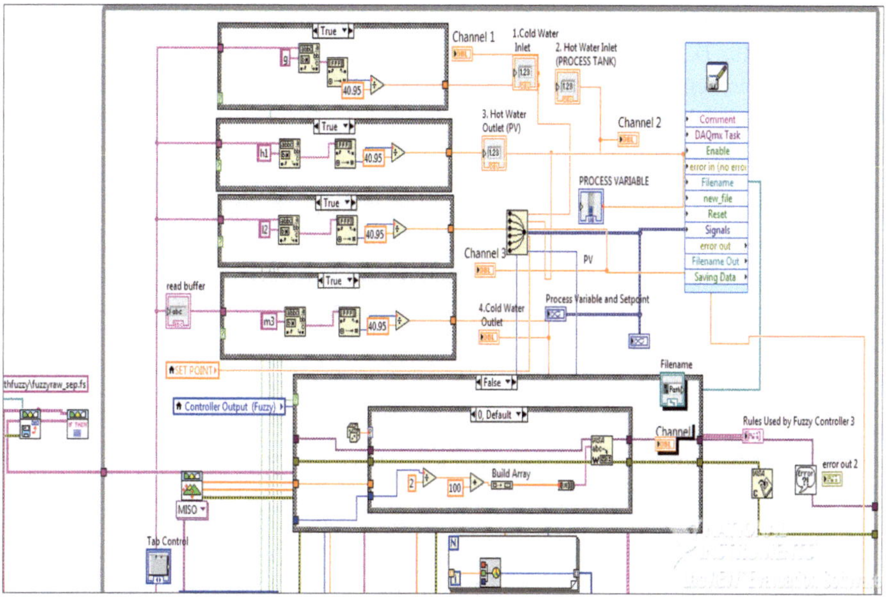

Fig. 11 LabVIEW data acquisition logic with RS232 serial communication

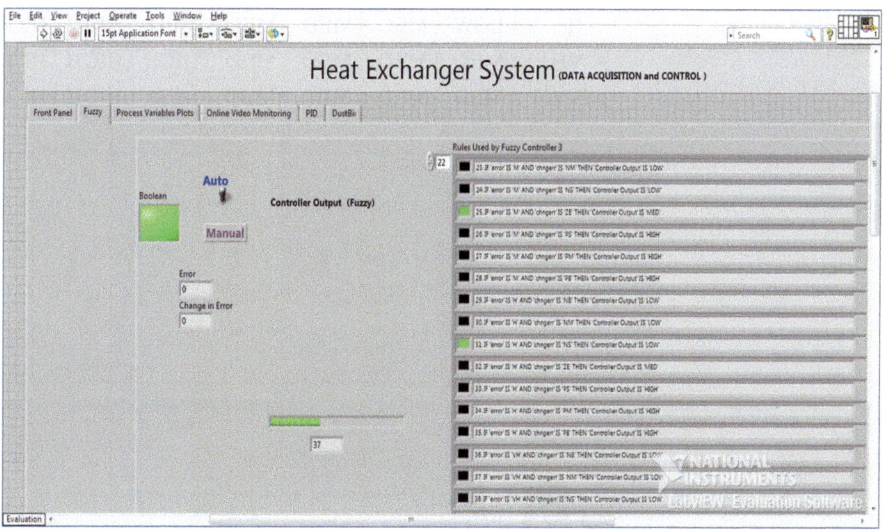

Fig. 12 Fuzzy rules display invoking '.fis' file in front panel for continuous monitoring of the process

bership function the universe of discourse for error, change in error and output are (−50 50), (−0.1 0.1) and (4 20) respectively.

Table 2 represents the formulation of rules with the abovementioned universe of discourse. Figure 13 describes fuzzy rules in the .fis file of the fuzzy system editor of fuzzy logic controller design.

In closed loop real time simulation Fuzzy controller gave response which is satisfactory. The real world STHE is shown in Fig. 14. A user-friendly graphical interface is developed as shown in Fig. 15 which controls the process. In the following section proposed controller performance is compared with Skogated's PI(D) controller. For this reason conventional controller was also implemented with FPGA based system interface.

Table 2 Fuzzy rule table for creating fuzzy associative memory

err	Change err						
	NB	NM	NS	ZE	PS	PM	PB
	NB	NM	NS	ZE	PS	PM	PB
VVL	LOW	LOW	LOW	LOW	HIGH	HIGH	HIGH
VL	LOW	LOW	LOW	LOW	HIGH	HIGH	HIGH
L	LOW	LOW	LOW	MED	HIGH	HIGH	HIGH
M	LOW	LOW	MED	MED	HIGH	HIGH	HIGH
H	LOW	MED	MED	MED	MED	HIGH	HIGH
VH	MED	MED	MED	MED	HIGH	HIGH	HIGH
VVH	MED	MED	MED	MED	MED	HIGH	HIGH

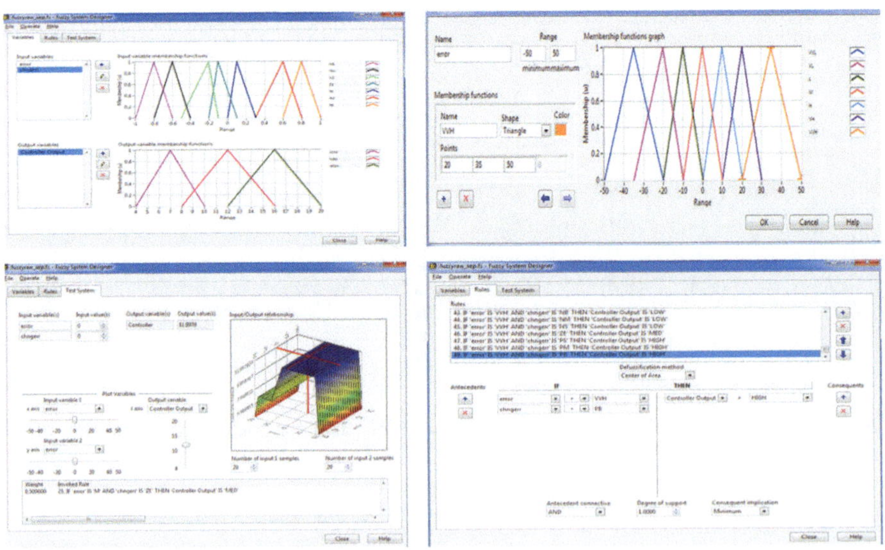

Fig. 13 The fuzzification and defuzzification procedure in fuzzy system editor window of LabVIEW

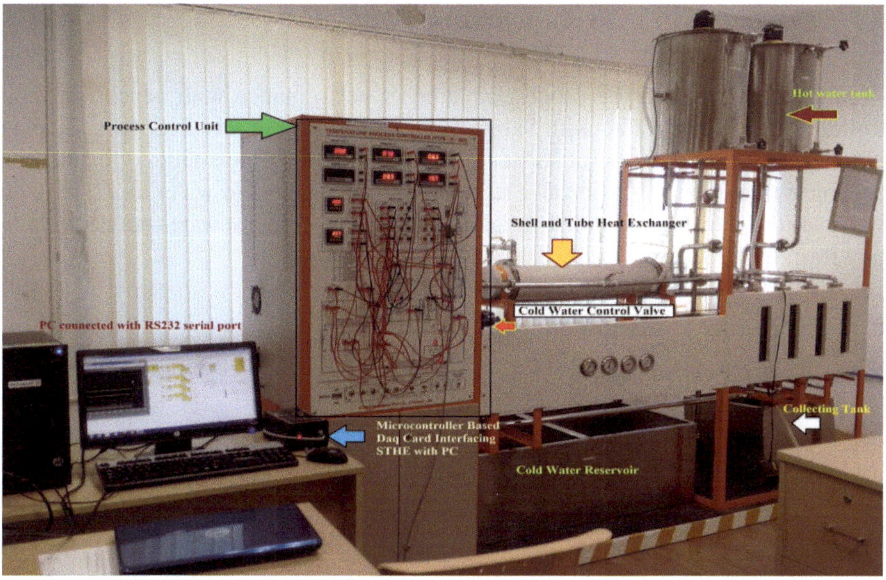

Fig. 14 Real time STHE process interfaced with DAQ card

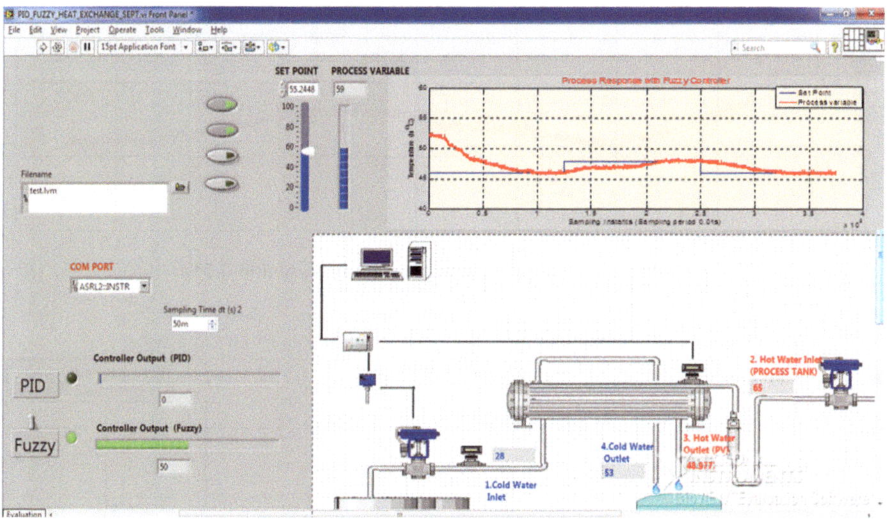

Fig. 15 Workflow schematics with fuzzy controller and plant

5.2 Comparison with Benchmark Controller

In order to distinguish the proposed method's performance with benchmark controller, STHE is controlled with Skogestad's PID controller (see Fig. 16). The plant is run using similar set point for temperature from hot water outlet. Opening and closing action of air-to-close valve regulates the inflow of cold water to STHE. Cold water inflow is the manipulated variable to get the desired hot water outlet temperature from the process.

5.3 Results Analysis

It is noticed in real time implementation that the process variable in STHE tracks the set-point temperature, however the proposed controller performed better than the

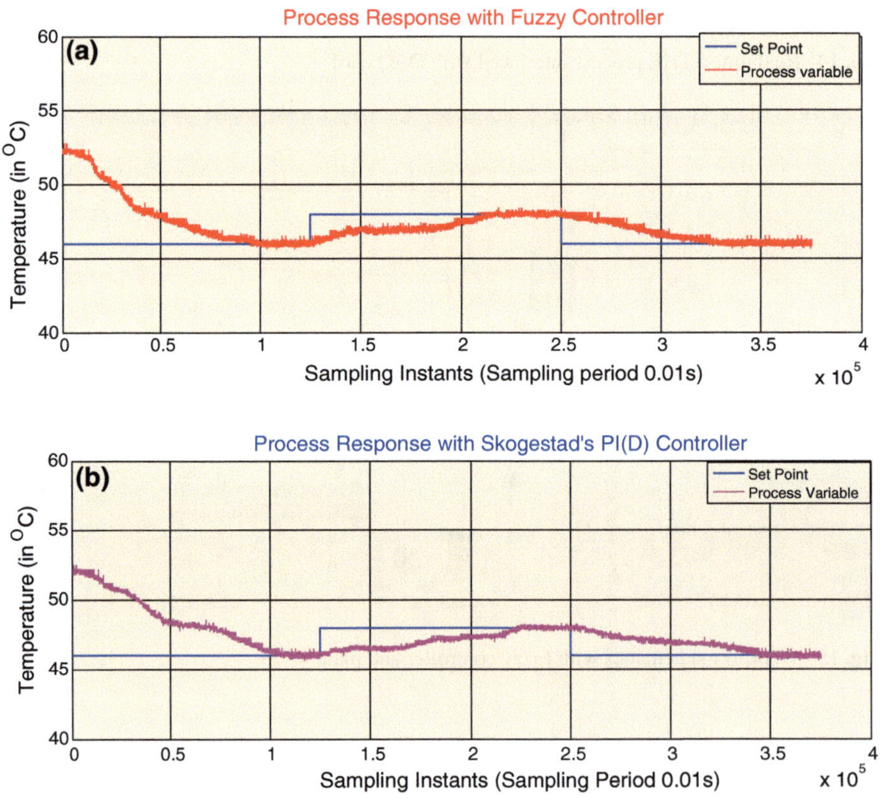

Fig. 16 **a** Real-time response of the fuzzy controller, and **b** Skogestad's PI(D) controller for the STHE using MPSoC based FPGA

(a)

(b)

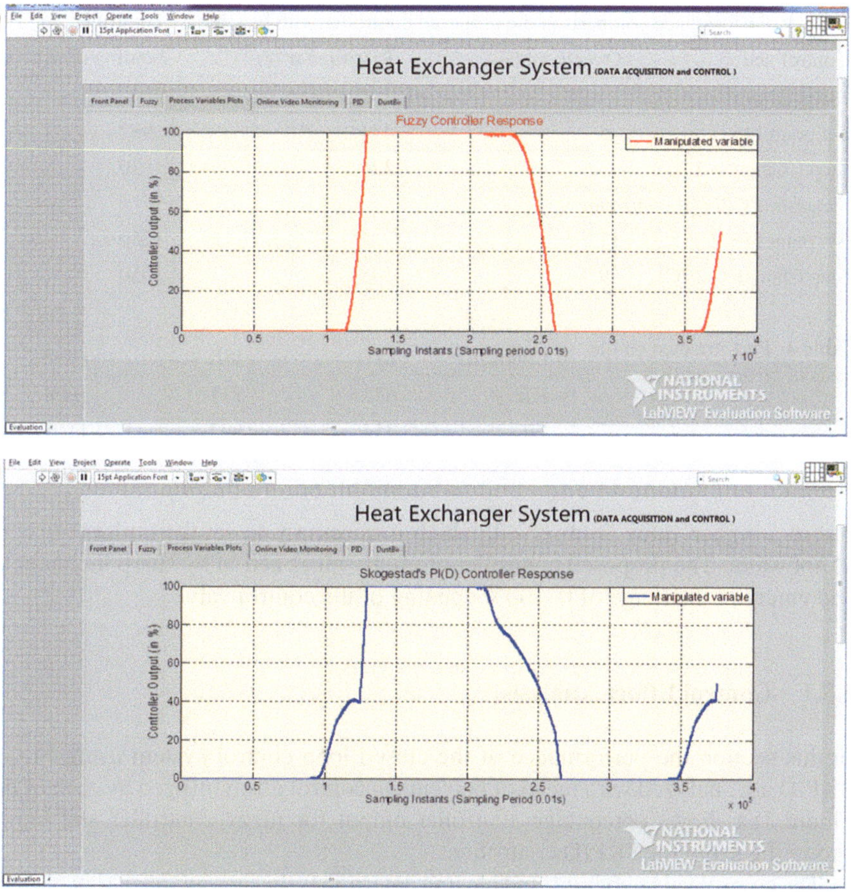

Fig. 17 **a** Fuzzy controller output and **b** Skogestad's PID controller output

PID controller. It is observed in Fig. 17 that the settling time for fuzzy controller with the set point tracking at 46 °C is 0.98 $\times 10^5$ sampling instants with the second change of 48 °C was 2.08 $\times 10^5$ which equals to 0.83 $\times 10^5$ samples from the sample at which the set point is changed and third change of 46 °C is tracked in 3.29 $\times 10^5$ sampling instants which equals to 0.79 $\times 10^5$ samples as from the corresponding set point change. However for the case of PID controller the first set point tracking occurred at 1.15 $\times 10^5$ sampling instants, second tracking for 2 °C rise in set-point happened at 2.2 $\times 10^5$ sampling instants and third change of set-point was reached at 3.41 $\times$ 10^5 sampling instants. Summary of performance of different controllers is presented in Table 3. Also performance index is calculated based on IAE, ISE and ITAE values which are shown in Table 4.

In Fig. 17 two controllers output is shown. It is noticed in real time that both controllers tracks set point temperature for STHE process. Cold water temperature

Table 3 Performance comparison the proposed controller with benchmark controller

Control action	Overshoot (%)	Rise time tr (s)	Settling time ts (s)
Fuzzy controller			
Set point (46 °C)	5	434	987
Step Change (+2 °C)	3	376	830
Skogestad's PI (D) controller			
Set point (46 °C)	11	722	1250
Step Change (+2 °C)	9	548	1050

Table 4 Performance on the basis of error reduction

Controller	IAE	ISE	ITAE
Fuzzy	677	811	1008
PID	754	892	1112

rise to a higher value is observed, which is necessary to set the outlet temperature of hot water. The proposed controller promptly takes action by sharp increment and decrement in the range of 0–100% opening of the control valve.

5.3.1 Control Effort Analysis

In this section the performance of the closed loop control system using FLC and SKPID are analyzed with mean of percentage control effort utilized by each of them. Figure 18a shows percentage controller output for fuzzy controller and Fig. 18b shows the same for SKPID controller.

The control valve opening and closing determines the amount of energy consumed i.e. the volumetric inflow of cold water to the shell side of STHE for regulation for hot water outlet temperature. In order to find the percentage controller effort the mean value has been calculated. There are 370,000 samples when the real time process is performed. Instantaneous percentage controller output has been accumulated over this period and the mean of this over the entire duration is considered as a measure of effective percentage control action as given by (1).

$$\text{Mean Control Effort} = \text{Sum of instantaneous control effort/duration} = \frac{\sum_{i=1}^{n} Y_i}{\text{No of Samples}}. \quad (1)$$

The computed mean percentage control effort for MPSoC based FLC and Skogestad's PID controller has been listed in Table 5. From the obtained results it can be concluded that the mean control effort with FLC is less compared to that with PID controller. This reflects lesser cold water inlet to the STHE with similar experimental condition with FLC controller compared to PID controller.

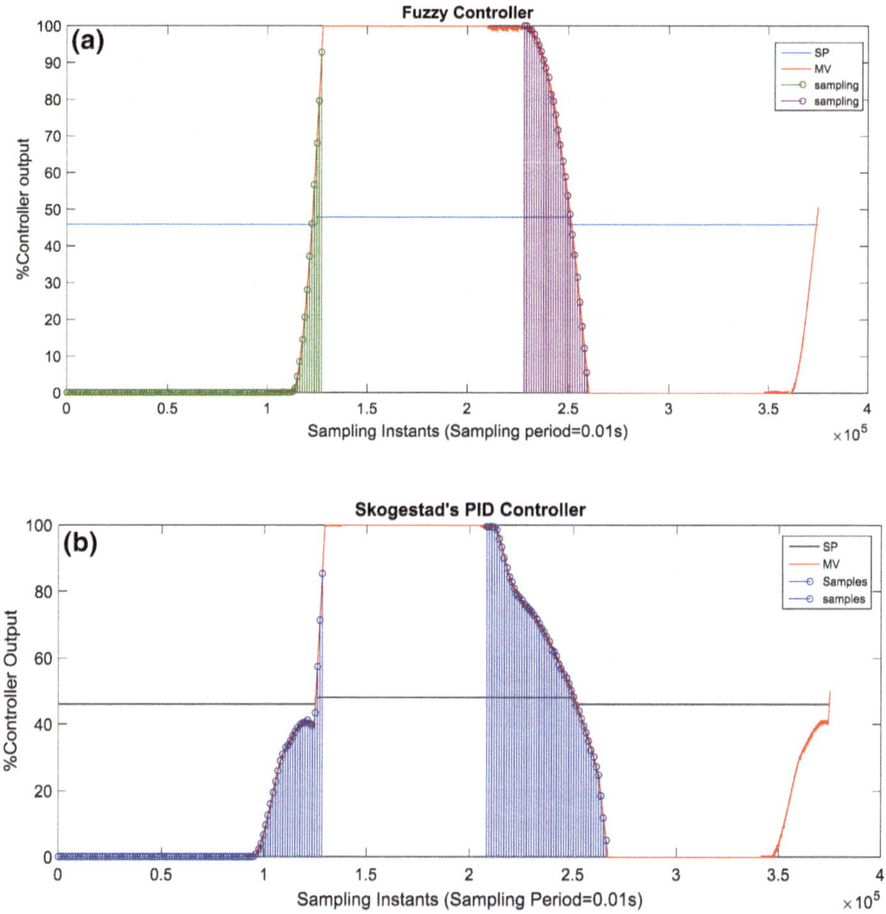

Fig. 18 **a** FLC effort on each set point change and **b** SKPID results compared on the same instants of change

Table 5 Percentage controller effort analysis

Controller on chip design	Mean value of the % age control effort
FLC on MPSoC	45.3562
SKPID on MPSoC	47.3734

6 Conclusion

In this work Fuzzy logic controller has been implemented successfully in real time on MPSoC based FPGA for STHE. This proposed work is a significant contribution to embedded control of real-time processes. The methodology provides a compact, flexible, fast and cost-effective control of process plants.

To implement, we used NI LabVIEW RT Software for configuring NI CRIO 9012 and CRIO9101. For this design, controller logic is kept embedded in FPGA and real time processor of CRIO 9012 communicates with RS232 utilizing FPGA scan interface mode. RS232 communicates with I/O modules and PCI bus interconnects FPGA for parallel processing of monitoring and control. FPGA acts as a commander to the plant to get desired temperature output.

The primary focus of SoC verification is to check the integration between various components, rather than implementing individual functionalities separately. We achieved target by running parallel loops in the software program and configured the FPGA cells to make it work as a system on chip. For this process we utilized predefined Intellectual Property (IP) library functions along with newly designed logic. To meet the objective of SoC we integrated all functionalities onto a chip and processes of acquisition, monitoring and control happened in parallel. On the basis of performance parameters effectiveness of the proposed algorithm is also justified. With tracking of set point and steady response, the proposed intelligent controller is shown to produce effective response based on settling time, ISE, IAE and ITAE values with minimum overshoot.

Acknowledgements The authors acknowledge Manipal Academy of Higher Education, Manipal, Udupi 576104, Karnataka, India for providing necessary infrastructure and financial assistance for carrying out this research work.

References

1. Huang H-C, Tsai C-C (2009, May) FPGA implementation of an embedded robust adaptive controller for autonomous omnidirectional mobile platform. IEEE Trans Ind Electron 56(5)
2. Jung SM, Lim MS (2007) System on chip design of embedded controller for car black box. In: International symposium on information technology convergence. IEEE. https://doi.org/10.1109/isitc.2007.68
3. Youness H, Moness M (2014, November) MPSoCs and multicore microcontrollers for embedded PID control: a detailed study. IEEE Trans Ind Inform 10(4)
4. Sun K, Wang M, Shao Z, Liu H, Wei H, Wang T (2010) Design and synthesis of a multiprocessor system-on-chip architecture for real-time biomedical signal processing in gamma cameras. J Signal Process Syst Springer Publ 59:71–83
5. Othman SB, Salem AKB, Abdelkrim H, Saoud SB (2012) MPSoC design approach of FPGA-based controller for induction motor drive. In: IEEE conference on industrial technology, March 19–21, Athens
6. Zhang Y, Guo X (2014) Development of intelligent traffic control system based on FPGA and single chip microcomputer technology. J Chem Pharm Res 6(1):1156–1159

7. Monmasson E, Idkhajine L (2011) FPGAs in industrial control applications. IEEE Trans Industr Inf 7(2):224–243
8. Rajesvari R, Manoj G, Angelin Ponrani M (2013) System-on-chip (SoC) for tele command system design. Int J Adv Res Comput Commun Eng 2(3):1580–1585
9. Ayala David JL, Alonso A, Reis R (2012) VLSI-SoC: forward-looking trends in IC and systems design. In: 18th IFIP WG 10.5/IEEE international conference on very large scale integration, VLSI-SoC, September 27–29, Madrid, Spain
10. Paul R, Chokkadi S (2016) Implementation of NARMA-L2 controller for shell and tube heat exchanger temperature process. Ind Eng Chem Res (I&ECR) 55:5644–5653 American Chemical Society Publications
11. Paul R, Shreesha C, Prabhu S, Shinde S (2015, December) Design of intelligent controller for temperature process on FPGA. Int J Control Theory Appl (IJCTA) 8(3):1113–1120. International Science Press
12. Paul R, Shreesha C, Shinde S (2015) LabVIEW implementation of fuzzy logic controller for heat exchanger process. In: IEEE international conference on futuristic trend in computational analysis & knowledge management (ABLAZE 2015), Noida, February 8–10. IEEE Xplore, pp 13–17. 978-1-4799-8433-6/15
13. Lin S (2009) The controller design based on fieldbus and system-on-chip. In: 2009 international conference on information management, innovation management and industrial engineering. IEEE. https://doi.org/10.1109/iciii.2009.447
14. Jha HR, Priyadarshi A (2012) System on chip (soc) architecture in industrial control system of hydraulic damper test bench using labview. Int J Eng Sci 1(9):28–32
15. Alujmah A, Ahmed MA (2015) Design of high speed data transfer direct memory access controller for system on chip based embedded products. J Appl Sci 15(3)
16. Padole D, Bajaj P (2008) Fuzzy arbiter based multi core system-on-chip integrated controller for automotive systems: a design approach. In: IEEE conference on automated system design, 5–7 May, Niagara Falls, Canada
17. Othman SB, Ghrissi M, Salem AKB, Soud SB (2008) FPGA hard core single processor implementation of RT control applications. In: International conference on design & technology of integrated systems in nanoscale era, 25–27 May, Tunisia
18. Economakos C, Kiokes G, Economakos G (2015) Using advanced FPGA SoC technologies for the design of industrial control applications. In: IEEE international conference on information, intelligence, systems and applications, 6–8 July, Corfu, Greece

Simultaneous Exploration and Coverage by a Mobile Robot

P. M. Mohammad Minhaz Falaki, Akshar Padman, Vishnu G. Nair and K. R. Guruprasad

Abstract In this paper, we propose a problem of simultaneous exploration and coverage for a mobile robot, combining the problems of area coverage with exploration and mapping. The primary task here is to completely cover an initially unknown region. Here we combine the advantages of online and off-line coverage path planning algorithms by using the exploration as an aid. The robots perform intermittent exploration during coverage in order to update the map of the environment, which in turn is used to generate the coverage path. We illustrate and demonstrate the problem using the off-line version of Spanning Tree Coverage algorithm with a frontier-based exploration strategy. The simulation results demonstrate that the robot successfully achieves complete and non-repetitive coverage.

1 Introduction

Problems such as area coverage, exploration and mapping along with localization are some of the very useful problems addressed in the mobile robot motion planning literature. Any motion planning, including point to point motion, area coverage, or exploration, typically requires localization and the map of the workspace. The problems of mapping (using exploration) and localization are dependent in the sense that mapping requires exact localization of the robot, while availability of the map

P. M. Mohammad Minhaz Falaki
Cellprop Pvt Ltd., Bengaluru, India

A. Padman · K. R. Guruprasad
Department of Mechanical Engineering, National Institute of Technology Surathkal, Surathkal, Karnataka, India

V. G. Nair (✉)
Department of Aeronautical and Automobile Engineering, Manipal Institute of Technology
Manipal Academy of Higher Education, Manipal 576104, Karnataka, India
e-mail: vishnu.nair@manipal.edu

C. Shreesha and R. D. Gudi (eds.), *Control Instrumentation Systems*,
Lecture Notes in Electrical Engineering 581,
https://doi.org/10.1007/978-981-13-9419-5_3

of the environment can aid in exact localization. Thus Simultaneous localization and mapping (SLAM) algorithms have been very popular in mobile robotic research community.

The terms exploration and coverage are used in the literature to refer to different related problems. A *coverage path planning* (*CPP*) algorithm has to generate a path such that a coverage tool attached to the robot move through each and every point (or cell in a gridded region) to perform certain task, such as vacuum cleaning, land-mine detection, lawn mowing, etc. In this paper, we refer to such problems as the *coverage* problem. Apart from this completeness of coverage, it is also desirable to have non-repetitive coverage, that is, the robot should not visit a point more than once. Off-line CPP algorithms use the map of the environment to generate a coverage path. This map may be obtained through an exploration process or the region is completely known a priori. Online CPP algorithms use onboard sensors to detect obstacles in order to generate an obstacle free coverage path. In contrast, by *exploration* we refer to the problem gathering of information on presence or absence of obstacles, in an area of interest, by a mobile robot equipped with necessary sensors. This information is typically used to obtain an occupancy (by obstacles) *map*. The purpose of map is to find the free areas within the region of interest where a mobile robot performing certain task can move freely. Further, most exploration strategies use a discretized space, and exploration is the process of identifying the nature of each cells, occupied or free, using the onboard sensors. The cells outside the sensor range are considered as unknown or unexplored.

Several algorithms have been proposed in the literature to solve the area coverage problems. Choset [1] provides a survey of several such CPP algorithms. A more recent survey is provided in [2]. Coverage algorithms may generate coverage path avoiding obstacles off-line (such as in [3] or off-line STC in [4]), when the map of the environment is known a priori, otherwise the robot uses onboard sensors to sense obstacles as and when they are encountered to generate path on the go (online) avoiding the obstacles (such as in [1] or online STC provided in [4]). Shnaps and Rimon [5] propose a CPP that generates optimal coverage path for battery powered mobile robots so as to reduce the battery usage. Acar and Choset [6], discuss sensor-based Morse decomposition for boustrophedon-like CPP. Detection of critical points is crucial step in Morse/Boustrphedon decomposition, and is affected by the sensor noise. In this paper the authors address the problem of use of imperfect sensors in unstructured environments and provide strategies for rejection of bad sensor data to improve the accuracy of detection of critical points.

In an *exploration* problem [7–11], the robot chooses an optimal point from where the exploration results in maximal information gain in terms of the map of the environment. The robot plans the path accordingly and performs exploration, typically using a long-range sensor to obtain a complete map of the environment.

While a long-range sensor is useful an exploration/mapping applications, short-range sensors are used in CPP problems. A mobile robot may be equipped with several senors which may include both long-range and short-range obstacle detection sensors. However, CPP algorithms either use only a short-range sensor in spite of availability of of long-range sensor, or may use only short-range information from the longer range sensors (such as say, ultrasonic or IR sensors) discarding the long-range information, as they need to detect an obstacle only in the next cell. In a typical online CPP algorithm, a robot may have to stop at each cell, sense the environment, and plan/decide on which of the neighboring cell to move next, and then move (act). However, in the case of an off-line CPP algorithm, entire path is planned at one go (no sensing required as the map is available a priori), and the robot may follow the path smoothly, and without any unnecessary stops.

The coverage algorithms reported in the literature are either off-line, using complete a priori knowledge about the arena, or online, using no a priori knowledge. SLAM algorithms may be as aid to both off-line and online CPP algorithms to improve robot localization. While off-line CPP algorithms have advantages of faster and smoother coverage path due to the fact that the 'sense' (the environment for obstacles) and 'plan' (the path covering the region avoiding the obstacle) phases are decoupled from the 'act' (actually move along the path generated/planned), they require complete map of the environment. Though the CPP algorithms do not require the map of the environment, robot has to perform 'sense' (the obstacles), 'plan' (the coverage path avoiding obstacles, and 'act' (moving along the planned path) at each cell within the region, leading to slower coverage and larger energy requirement.

Motivated by SLAM problem where two different but related problems of localization and mapping are solved together, where localization aids in mapping and mapping improves localization, in this paper, we propose a Simultaneous Exploration and Coverage (SimExCoverage) problem to combine the advantages of both online and off-line CPP. Here, the (off-line) CPP strategy can utilize available partial knowledge about arena (in terms of map), while by intermittent exploration the robot updates this knowledge on the go. Though the primary objective here is to cover the arena, a map of region can be obtained at the end as a byproduct. Though conceptually any robot coverage algorithm (such STC [4], Boustrophedon [1], etc.) and any exploration and mapping algorithm, can be used for solving the proposed SimExCoverage problem, the algorithm that solves the problem still depends on the underlying CPP and exploration algorithm. In this paper, we use the off-line STC [4] and a frontier- based exploration strategy as the underlying CPP algorithm and exploration algorithm and solve the SimExCoverage problems.

2 Problem setting

We consider a convex region $Q \subset \mathbb{R}^2$ decomposed into $2D \times 2D$ square cells, as the area to be covered by a mobile robot whose coverage footprint (or the footprint of the coverage tool) is assumed to be $D \times D$ square. Each $2D \times 2D$ cell is called a major cell, which has four $D \times D$ minor cells. A major cell is either 'free' of obstacle, or 'occupied' by the obstacle. Thus, even if a major cell is partially occupied by an obstacle, it is assumed to be completely occupied. For this reason, Choset [1] classifies such algorithms as approximate cellular decomposition based CPP algorithms. The coverage is considered complete when the robot passes through all minor cells, and

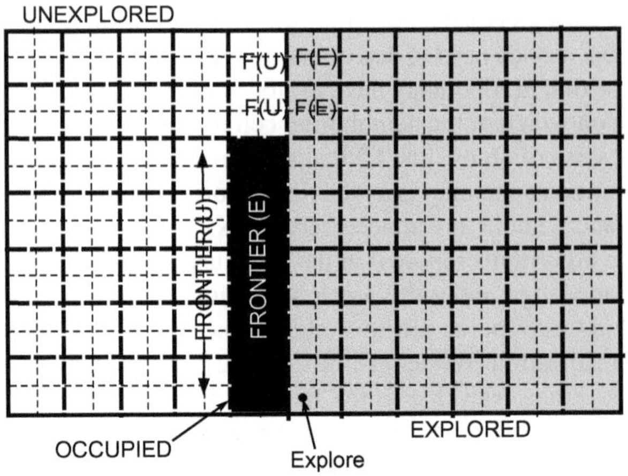

Fig. 1 The arena is decomposed into 'major cells' of size $2D \times 2D$ (shown with thick long dashed boundary) and each major cell has four $D \times D$ sized minor cells (shown with thin dashed line boundary). Dark cells are 'occupied' (by obstacles) and remaining cells are 'free' (of obstacle). An instance of exploration from divides the region into 'explored' and 'unexplored' regions. Explored region is made up of explored free cells (grey cells) explored occupied cells (dark cells). The frontier is set of explored ('FRONTIER(E) or F(E)) and unexplored ('FRONTIER(U) or F(U)) cells on the boundary separating the explored and unexplored regions

non-repetitive, if no minor cell is visited more than once. The problem setting is suitable for algorithms such as STC [4] used as the underlying CPP algorithm.

Figure 1 illustrates the problem setting discussed here. The exploration process divides the major cells into 'known' (or 'explored') and 'unknown' (or 'unexplored') cells. A known cell or explored cell may be 'free' or 'occupied'. In Fig. 1, explored 'free' cells are shaded gray and explored 'occupied' cells are shown with black. The unshaded (white) cells are unexplored or unknown cells. Frontier cells form the boundary between explored and unexplored regions. The frontier cells may be 'explored frontier' (part of explored region) or 'unexplored frontier' (part of unexplored region) cells. The 'explored' frontier cells are marked 'FRONTIER(E)' or 'F(E), while the 'unexplored frontier cells are marked 'FRONTIER(U)' or 'F(U) in the figure. When a free F(E) cell has an adjacent free F(U) cell, an exploration window is formed. That is, a robot located in this free F(E) cell can access the unexplored region through the adjacent free F(U) cell. We call an adjacent pair of free F(E) and F(U) cells as an *exploration window*. In Fig. 1, two pairs of cells marked F(U) and F(E) are two exploration windows.

In this work, we assume that the robot is equipped with a long-range exploration sensor such as LIDAR. The CPP algorithm uses the map provided by the exploration algorithm, and hence does not make use of any sensors. We assume that the robot is exactly localized, as the main focus of this work is to propose and demonstrate a simultaneous exploration and coverage problem. SLAM or other localization techniques may be used in practice.

3 Proposed Simultaneous Exploration and Coverage Methodology

In the proposed Simultaneous exploration and coverage problem, we combine the exploration and mapping and the CPP problems, the primary task being area coverage. Figure 2 illustrate the simultaneous exploration and coverage problem. Exploration generates the map, which the CPP algorithm uses to generate the coverage path. The robot moves along the coverage path. While performing coverage, at certain points along the coverage path, robot performs exploration. Unlike in an exploration problem, exploration strategy in SimExCoverage does not perform path planning. Localization is very important for any path planning algorithm. In this work, we assume exact localization is available as the focus of the work is on combining exploration and coverage problems to generate complete and non-repetitive coverage path. In practical applications, techniques such as SLAM may be used to obtain more precise localization of the robots. In fact, exploration and mapping are part of the simultaneous exploration and coverage problem addressed here and hence incorporating SLAM algorithms is not very difficult.

Before the beginning of the first exploration, all (major) cells are considered ‘unknown’ or ‘unexplored’, while the ‘explored’ and ‘frontier’ lists are empty. In this paper we use off-line version STC algorithm [4] for CPP of the explored region. A minimal spanning tree (MST) using Kruskal’s algorithm is created over the graph formed by major cells within the explored region, using the list of ‘free’ cells. As the robot moves along the CP, whenever it is on a ‘frontier’ (explored) cell and it finds an ‘exploration window’, the robot switches to ‘explore’ mode. A MST is created when a newly ‘explored’ region is added, and is appended to the existing ST, and the CP is continued from the current cell. Each exploration process involves a full scan using the exploration sensor (LIDAR) and updating the lists of ‘free’, ‘occupied’ cells and the ‘frontier’ cells.

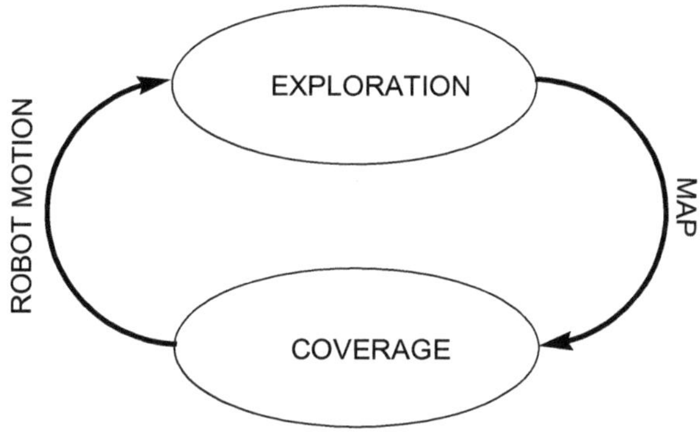

Fig. 2 SimExCoverage problem combines Exploration and Coverage (CPP) problems. Exploration provides map for coverage path planning, while the CPP provides the path for robot motion. While the robot is moving along the coverage path, at exploration is carried out at a certain locations

4 Results and discussions

Now we shall describe the Simultaneous exploration and coverage-STC using a simulation result as an illustrative example. The simulations have been carried out using Matlab to generate the coverage path using the proposed SimExCoverage strategy with off-line STC and frontier-based explorations strategies described int he previous sections as the underlying CPP and exploration algorithms. We consider a scenario as shown in Fig. 3. The region is decomposed into 6×6 major cells. The cells $7, 9, 12, 13, 14, 15, 17$, and 18 are occupied (by obstacle) and rest of the cells are free. Initially the robot is located in a minor cell of the major cell no. 19.

The robot performs an exploration from cell no. 19. The result of this exploration is shown in Fig. 4a. Explored 'free' cells are shown with white. Explored 'occupied' cells are 13, 14, 15, 17, and 18. The remaining region (cells) are unexplored. Now a ST is created in this explored region, shown by red lines passing through major nodes (red dot). The corresponding CP is created as the robot moves along this path (shown in blue line), as shown in Fig. 4b. When the robot reaches a minor cell in the major cell 23, which is a explored 'frontier' cell, it shares boundary with the unexplored 'frontier' cell 16, it performs second exploration. With the second exploration, few more cells are are explored (shown in white), and a ST is created through the newly added 'explored' cells. This is illustrated in Fig. 4c. Now when the robot reaches a minor cell in the major cell 4 (see Fig. 4c), an explored 'frontier' cell, robot performs exploration as this cell shares a boundary with unexplored 'frontier' cell 3. The ST is created in this newly explored region (cells 1, 2, and 3). The robot CP is created as the robot moves on the right side of the ST edges. While the robot reaches a minor cell in the major cell 2 from major cell 1, it explores and finds that cell 8 is free, and a ST edge is added to this cell. Finally, the robot reaches the starting major cell 19 as shown in Fig. 4d. The next minor cell from this position is the same as the starting minor cell. Thus, the robot terminates Simultaneous exploration

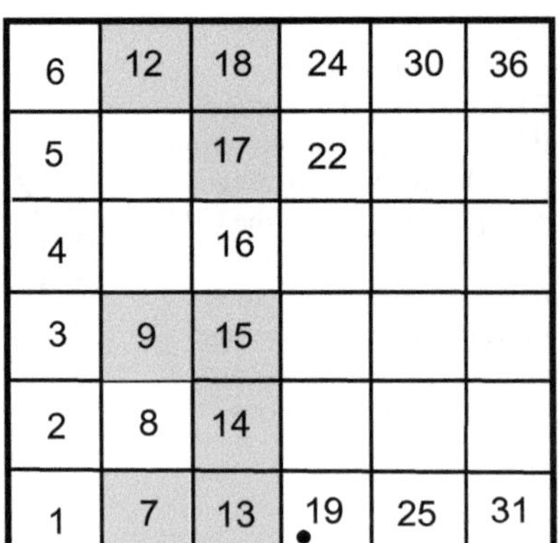

Fig. 3 The arena to be covered is decomposed into cells of size $2D \times 2D$. Cell numbering format is as shown. Occupied cells $(7, 9, 12, 13, 14, 15, 17$, and $18)$ are shaded. Robot starts at a minor cell in the major cell no. 19

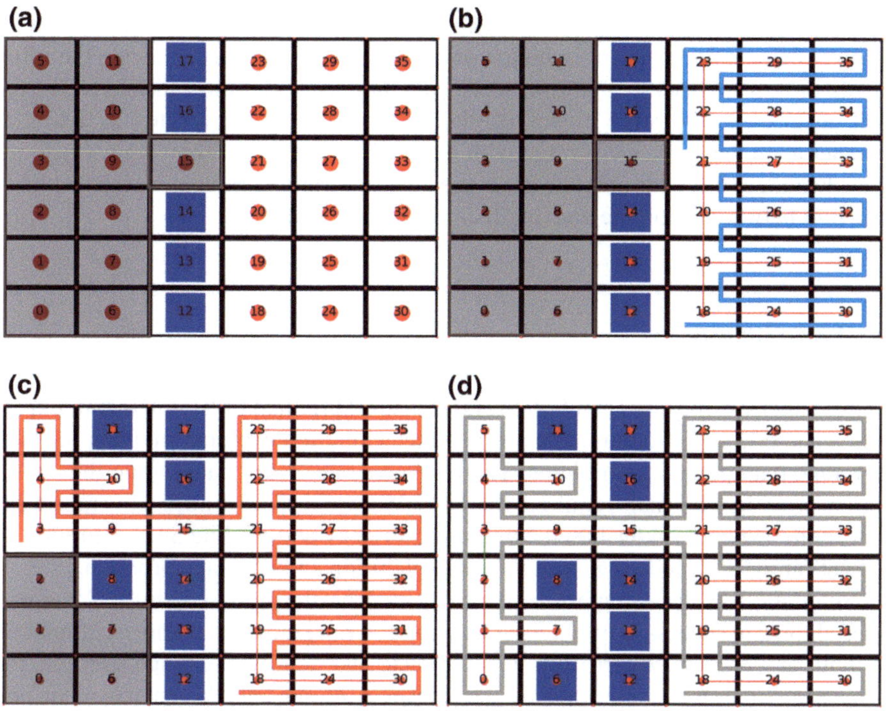

Fig. 4 Coverage path and exploration steps. **a** First exploration, **b** second exploration instance after complete coverage of free explored cells, **c** third exploration instance, and **d** the last exploration and successful completion of the coverage. Explored free (major) cells are unshaded, unexplored cells are shaded with gray, and explored 'occupied' cells are shaded with blue (dark gray/black without color). The nodes corresponding to major cells are shown with (red) dots. The ST edges are (red) lines passing through nodes corresponding to the free major cells. Coverage path (at graph-level, not actual robot path) passes through minor nodes/cells (not shown in Figure for clarity) around the ST edges is shown with blue lines

and coverage-STC. As we observe here at this instance, entire region is explored and coverage is complete and without any coverage overlap/repetitive coverage. Note that unlike the online STC algorithm (or any online CPP algorithm), here the robot does not use its sensors, which apart from requiring energy and time, also requires the robot (or the sensor alone) to scan all the neighbors (270°) at each minor cell. During coverage the robot has to simply follow the CP (as in off-line CPP), and only look for frontier cells, where the mode is switched to exploration. Also, unlike an off-line CPP algorithm, SimExCoverage-STC does not require a priori knowledge of the map of the arena. Apart from the complete and non-repetitive coverage, Simultaneous exploration and coverage-STC generates the map of the environment which may be used for any other purposes, including repeated area coverage. Any repeated area coverage can now be performed completely off-line.

Note that compared to the coverage using online STC algorithm which requires 28 number of sensing instances, with the proposed SimExCoverage strategy, the robot needs only 3 instances of sensing corresponding to exploration instances. In the rest of 25 cells, the robot moves continuously along the CP created in a similar manner as in the case of coverage with the off-line STC algorithm. This illustrates the advantages of combining the coverage and exploration problem which results in a hybrid between an online and an off-line coverage strategies.

5 Conclusions

In this paper we proposed SimExCoverage, a novel problem of 'simultaneous exploration and coverage' for a mobile robot, which combines exploration, mapping, and coverage path planning problems. The CPP generates robot path, while the exploration provides the map required for CPP. We proposed a SimExCoverage-STC algorithm using a frontier-based exploration strategy and off-line STC algorithm as a solution to the proposed Simultaneous exploration and coverage problem.

Simulation results at graph-level demonstrated that the proposed SimExCoverage-STC algorithm successfully covers arena, not known to the robot a priori, completely without any overlap. Apart from this complete and non-overlapping coverage, the proposed SimExCoverage reduces the instances of sensing the environment and planning considerably in comparison with the traditional online coverage strategy.

A detailed formal analysis of the proposed SimExCoverage strategy and experimental validation of the proposed strategy are part of ongoing work.

References

1. Choset H (2001) Coverage for robotics—a survey of recent results. Ann Math Artif Intell 31(1–4):113–126
2. Galceran E, Carreras Marc (2013) A survey on coverage path planning for robotics. Robot Auton Syst 61:1258–1276
3. Choset H (2000) Coverage of known spaces: the boustrophedon cellular decomposition. Auton Robot 9:247–253
4. Gabriely Y, Rimon E (2001) Spanning tree based coverage of continuous areas by a mobile robot. Ann Math Artif Intell 31:77–98
5. Shnaps I, Rimon E (2016) Online coverage of planar environments by a battery powered autonomous mobile robot. IEEE Trans Autom Sci Eng 13:425-436
6. Acar EU, Choset H (2002) Sensor-based coverage of unknown environments: incremental construction of morse decompositions. Int J Robot Res 21:345–366
7. Lumelsky S, Mukhopadhyay S, Sun K (1990) Dynamic path planning in sensor-based terrain acquisition. IEEE Trans Robot Autom 6(4)
8. Lee D, Recce M (1997) Quantitative evaluation of the exploration strategies of a mobile robot. Int J Robot Res 16(4):413–447

9. Yamauchi B (1998) Frontier-based exploration using multiple robots. In: Proceedings of the second international conference on autonomous agents, pp 47–53
10. Gonzalez-Banos HH, Latombe JC (1998) Planning robot motions for range-image acquisition and automatic 3D model construction. In: AAAI fall symposium
11. Albers S, Kursawe K, Schuierer S (1999) Exploring unknown environments with obstacles. In: Proceedings of the 10th symposium on discrete algorithms

Tracking Control and Deflection Suppression of an AMM Modelled TLFM Using Backstepping Based Adaptive SMC Technique

Kshetrimayum Lochan, Jay Prakash Singh and Binoy Krishna Roy

Abstract Manipulators are widely used in all areas of science and technology. Effective trajectory tracking and quick deflection suppression are the two main aspect of research for a flexible manipulator. The paper reports aperiodic signal like trajectory tracking control for a planar assumed modes modelled two-link flexible manipulator (TLFM). The aperiodic chaotic signal is used as a desired trajectory for the TLFM. Thus, designing of a robust controller for the aperiodic signal tracking control is a challenging task. A backstepping based adaptive SMC technique is designed for the considered problem. In adaptive SMC, the gain of the switching control law is estimated online. The effectiveness of the considered controller is compared to an available backstepping controller. It is found that the designed backstepping based adaptive SMC perform better in terms of smaller tracking time, quick tip deflection suppression and lesser, smoother control efforts. Proposed trajectory strategy is validated on a two-link flexible manipulator in MATLAB simulation environment.

Keywords Backstepping control · Adaptive SMC · Two-link flexible manipulator · Tracking control · Link deflection

Nomenclature

X_i, Y_i	Inertial frame axis
$\hat{X}_i, \hat{Y}_i$	Axis of rigid body moving frame
M_p	Payload mass
q	Vector of generalised coordinate of the manipulator

K. Lochan (✉)
Department of Mechatronics, Manipal Institute of Technology, Manipal 576104, Karnataka, India
e-mail: lochan.nits@gmail.com

J. P. Singh
Department of EE, Rewa Engineering College, Rewa 486002, MP, India

B. K. Roy
National Institute of Technology Silchar, Silchar 788010, Assam, India

C. Shreesha and R. D. Gudi (eds.), *Control Instrumentation Systems*,
Lecture Notes in Electrical Engineering 581,
https://doi.org/10.1007/978-981-13-9419-5_4

τ_i	Actuated torque input
j	No. of modes for the link i
ϕ_{ij}	Assumed spatial mode shapes
δ_{ij}	Time varying variables associated with the ϕ_{ij}
l_i	Length of ith link
i	No. of links

1 Introduction

Recent research in the field of robot manipulator is divided into rigid and flexible. Flexible manipulators are relatively more attractive at the application point of view. Some of the applications in the fields includes space, industries, hospitals etc., as compared with rigid counterparts [1, 2]. In flexible, the manipulators are classified mainly into flexible link and flexible joints manipulators. Another category is flexible link with flexible joint manipulator. Flexible link manipulators are more considered for study as compared with other types of flexible manipulators. It has various type based on the number of link and flexible nature like 1-link, two-link and multi-link flexible manipulator. Two-link flexible manipulators (TLFMs) are more considered for the research as compared with other types of flexible manipulator [3, 4]. Various advantages of the flexible manipulators create issues like non-minimum phase, nonlinearity, under actuation, noncolocation, etc. [5]. Tracking control and vibration suppression are the two main most acceptable and challenging control problem for any FMs as compared with other control problem like position control [6].

A physical TLFM model in its mathematical form using some modelling methods is used. The types of method considered for the modelling decide the effectiveness of the performance of the FM. The modelling of a TLFMs are obtained using the lumped parameter method, finite element method (FEM) or assumed modes method (AMM) [2]. Assumed modes method is more attractive among the modelling methods [7]. Various controllers are available in the literature for the tracking control of a TLFMs like adaptive control [1, 8], fuzzy logic control [6], H_∞ control [9], model predictive control [4], SMC [10–12], observer-based [13], LMI based PD control [14], etc. Among these SMC is considered as more better in terms of fast and robust tracking as well as for quick deflection suppression [15]. Many variant of SMC are being designed for TLFM like adaptive SMC [16], integral SMC [11], fuzzy SMC [17], neural SMC [18], second order SMC [15, 19], non-singular terminal SMC [20], second order terminal SMC [20], etc. Backstepping control is an efficient and effective control technique in the presence of disturbances [21]. Backstepping control is also being used for different control problems of a rigid manipulator [22–28]. Tracking control for a single link flexible joint manipulators using backstepping controller is discussed in [29–32]. Backstepping control is also used for single link FM [33–35]. The work in [13] discussed the tracking control of a two-link flexible space manipulator using backstepping with extended state observer. But, designing of robust control technique like SMC together with backstepping is very less explored in the literature [36] and is still a challenging task. Based on the above discussions, this paper motivated to design a backstepping based adaptive sliding mode controller for the chaotic trajectory tracking control of a TLFM.

The contribution in terms of work done in the paper is discussed in the following paragraph. In the paper, a backstepping based adaptive SMC is deliberated for tracking control of an AMM modelled planner two-link flexible manipulator. Here aperiodic chaotic signal is used as desired trajectory for the TLFM. Thus, tracking of an aperiodic signal as desired trajectory and quick deflection suppression is a challenging task. It is shown that the designed backstepping based adaptive SMC perform better in terms of faster trajectory tracking, deflection suppression for the considered TLFM is accomplished efficiently with negligible small control efforts. These performances are shown in comparison with an available backstepping controller in [37].

The remaining paper goes like this. Section 2 discussed the problem statement of the paper. The dynamic of the considered TLFM is presented in Sect. 3. The designing of the backstepping control and backstepping based adaptive SMC for a TLFM is given in Sect. 4. Section 5 presents the discussion along with the results of the controller designed. And finally, the paper is concluded in Sect. 6.

2 Problem Statement-Tracking Control Using Backstepping Based Adaptive SMC Technique

Consider the dynamics of a flexible manipulator is written as

$$\begin{cases} \dot{x}_1 = x_2 \\ \dot{x}_2 = f(x_1, x_2) + b\tau \end{cases} \tag{1}$$

where x_1 and x_2 are the states vector containing joint angles and mode shapes, $f(x_1, x_2)$ is the nonlinear function of various elements consisting of dynamics, stiffness and damping matrices, and b is the coupling term for the input τ.

Here first the procedure of designing backstepping controller for the system given in Eq. (1) is presented, then procedure of designing adaptive sliding mode controller is presented.

Step 1: In this step an error variable is defined for the tracking of desired trajectory using the difference between the desired trajectory and the first state variable. Consider the error variable is $e_1 = x_{1d} - x_1$, then its derivative can be written as $\dot{e}_1 = \dot{x}_{1d} - x_2$. A candidate Lyapunov function is considered and its derivative is given as

$$\begin{cases} v_{1p} = \frac{1}{2}e_1^2 \\ \dot{v}_{1p} = e_1\dot{e}_1 \\ \dot{v}_{1p} = e_1(\dot{x}_{1d} - x_2) \end{cases} \tag{2}$$

Here $\dot{v}_{1p}$ is negative definite if the desired value of x_2 is considered as $x_{2d} = \alpha_1(x_1) = \dot{x}_{1d} + k_1 e_1$, then the difference between the actual and the desired x_2 can be defined as

$$\{e_2 = x_{2d} - x_2 = \dot{x}_{1d} + k_1 e_1 - x_2 \tag{3}$$

Step 2: Using (3), the dynamics of error e_1 can be obtained as

$$\dot{e}_1 = \dot{x}_{1d} - x_2 = \dot{x}_{1d} - (\dot{x}_{1d} + k_1 e_1 - e_2) = -k_1 e_1 + e_2 \tag{4}$$

Using (1) and (3) the dynamics of e_2 is obtained as

$$\{\dot{e}_2 = \dot{x}_{2d} - \dot{x}_2 = \dot{x}_{2d} - (f(x_1, x_2) + b\tau) = k_1 - (f(x_1, x_2) + b\tau) \tag{5}$$

Now to show the stability of error dynamics e_1 and e_2 and to find the control input, another candidate Lyapunov function is considered as

$$\begin{cases} v_{2p} = \frac{1}{2}e_1^2 + \frac{1}{2}e_2^2 \\ \dot{v}_{2p} = e_1\dot{e}_1 + e_2\dot{e}_2 \end{cases} \tag{6}$$

Using the dynamics of e_1 and e_2, the derivative of Lyapunov function $\dot{v}_{2p}$ is written as

$$\begin{cases} \dot{v}_{2p} = e_1(-k_1 e_1 + e_2) + e_2(k_1 - (f(x_1, x_2) + b\tau)) \\ \dot{v}_{2p} = -k_1 e_1^2 + e_1 e_2 + e_2(k_1 - (f(x_1, x_2) + b\tau)) \end{cases} \tag{7}$$

In (7) $\dot{v}_{2p}$ is negative definite if the control input τ is selected as

$$\tau = \frac{1}{b}(-f(x_1, x_2) + k_1 + e_1 - k_2 e_2) \tag{8}$$

Then the derivative of Lyapunov function $\left(\dot{v}_{2p}\right)$ is written as

$$\dot{v}_{2p} = -k_1 e_1^2 - k_2 e_2^2 \tag{9}$$

Therefore control input defined in (8) stabilise the system and helps in desired state tracking with the condition $k_1, k_2 > 0$.

To increase the accuracy in tracking of desired trajectory and quick suppression of tip deflection, the backstepping controller is modified to design backstepping adaptive sliding mode controller.

Backstepping adaptive SMC is designed at the final stage of the backstepping controller. This is designed using the error variables with the help of following sliding surface as

$$\sigma = k_3 e_1 + e_2 \tag{10}$$

where k_3 is positive real number and are selected such that the polynomial in (10) is Hurwirtz which guarantee a stable sliding surface. Stability of the sliding surface is shown by considering another Lyapunov function v_{3p} along with the error e_1.

$$v_{3p} = \frac{1}{2}e_1^2 + \frac{1}{2}\sigma^2 \tag{11}$$

Using the derivative of error (4) sliding surface (10), the derivative of Lyapunov function v_{3p} is written as

$$\begin{cases} \dot{v}_{3p} = e_1\dot{e}_1 + \sigma\dot{\sigma} \\ \quad = e_1(-k_1e_1 + e_2) + \sigma(k_3\dot{e}_1 + \dot{e}_2) \\ \quad = e_1(-k_1e_1 + e_2) + \sigma(k_3\dot{e}_1 + k_1 - (f(x_1, x_2) + b\tau)) \end{cases} \tag{12}$$

Now consider the control input τ as in (13),

$$\tau = \frac{1}{b}((f(x_1, x_2) + k_1 + k_3\dot{e}_1 + \sigma^{-1}e_1e_2 + \rho sign(\sigma)) \tag{13}$$

$\dot{v}_{3p}$ is negative definite as discussed in (14).

$$\dot{v}_{3p} \leq -k_1e_1^2 - \varphi|\sigma| \tag{14}$$

Thus, using the Lyapunov theorem it is seen from (14) that $\dot{v}_{3p}$ is negative definite. Hence, we can say that the tracking of desired trajectory for system is achieved.

Now suppose the gain of the corrective control law φ in (13) is unknown and is adaptively estimated with time, then the control input in (13) is modified as

$$\tau = \frac{1}{b}((f(x_1, x_2) + k_1 + k_3\dot{e}_1 + \sigma^{-1}e_1e_2 + \hat{\varphi}\, sign(\sigma)) \tag{15}$$

where $\hat{\varphi}$ is the estimate of φ. Adaptive mechanism for $\hat{\varphi}$ is considered as [38].

$$\dot{\hat{\varphi}} = \dot{\tilde{\varphi}} = Q^{-1}(|\sigma|) \tag{16}$$

where $Q > 0$ and $\tilde{\varphi} = \hat{\varphi} - \varphi$. The stability of the sliding surface with the new control law (13) can shown by considering another candidate Lyapunov function as in (17).

$$v_{4p} = \frac{1}{2}e_1^2 + \frac{1}{2}\sigma^2 + \frac{1}{2}Q\tilde{\varphi}^2 \tag{17}$$

Taking the time derivative of (17) and using (4), (10) and (16), it is written as

$$\begin{cases} \dot{v}_{4p} = e_1\dot{e}_1 + \sigma\dot{\sigma} + Q\tilde{\varphi}\dot{\tilde{\varphi}} \\ \dot{v}_{4p} = -k_1e_1^2 - \hat{\varphi}|\sigma| + Q^{-1}\tilde{\varphi}(|\sigma|) \\ \quad \le -k_1e_1^2 - \varphi|\sigma| \end{cases} \tag{18}$$

Since $\dot{v}_{4p} < 0$ for $\varphi > 0$, hence motion on the sliding surface (σ) is stable.

3 Dynamics of a Planar Two-Link Flexible Manipulator

The graphical view/representation of a planar two-link flexible manipulator is presented in Fig. 1.

Link deflection of the two-link flexible manipulator shown in Fig. 1 is represented by $u_i(x_i, t)$ and is defined as [39, 40]

$$u_i(x_i, t) = \sum_{j=1}^{n} \emptyset_{ij}(x_i)\delta_{ij}(t) \tag{19}$$

The tip position of the ith link of the two-link flexible manipulator can be written as [39, 40].

$$y_i = \frac{1}{l_i}[u_i(x_i, t)] \tag{20}$$

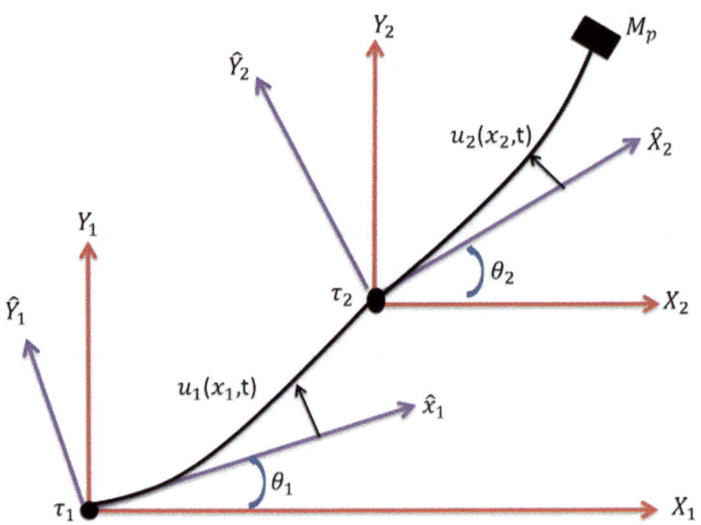

Fig. 1 The pictorial representation of a planar TLFM

Suppose the Lagrangian generalised coordinate are considered as $q = (\theta_i, \delta_{ij}(t)) \in R^{6\times 1}$ and by using the Lagrange AMM, the motion equation of a planner TLFM [5, 40] can be derived as

$$M(q)[\ddot{q}] + h(q, \dot{q}) + K[q] + D[\dot{q}] = \tau + \tau_d \tag{21}$$

where $M(q) \in R^{6\times 6}$ is the mass inertia matrix, $h(q) \in R^{6\times 1}$ is the centrifugal and coriolis force vector, $K(\delta_{ij}) \in R^{6\times 6}$ is the positive definite stiffness matrix, $D \in R^{6\times 6}$ is the positive definite damping matrix and $\tau \in R^{2\times 1}$ is joint torque vector.

4 Backstepping Controller Design

In this section, backstepping controller design method for a TLFM dynamics given in (21) is presented.

4.1 *Transformation of Variables*

Suppose the coordinates for the system (21) are defined in two new variables as $\theta_n = (\theta_1, \theta_2)^T$, $q_n = (\delta_{11}, \delta_{12}, \delta_{21}, \delta_{22})^T$. Using these variables the system matrices are written as [39, 40]:

$$M = \begin{bmatrix} M_{11} & M_{12} \\ M_{21} & M_{22} \end{bmatrix}, D = \begin{bmatrix} D_{11} & D_{12} \\ D_{21} & D_{22} \end{bmatrix}, H = \begin{bmatrix} H_1 \\ H_2 \end{bmatrix}, K = \begin{bmatrix} K_1 \\ K_2 \end{bmatrix} \tag{22}$$

Using the matrices defined in (22), the dynamics of TLFM given in (21) are written as

$$M_{11}\ddot{\theta}_n + M_{12}\ddot{q}_n + D_{11}\dot{\theta}_n + D_{12}\dot{q}_n + H_1 + K_1 = \tau \tag{23}$$

$$M_{21}\ddot{\theta}_n + M_{22}\ddot{q}_n + D_{21}\dot{\theta}_n + D_{22}\dot{q}_n + H_2 + K_2 = 0 \tag{24}$$

Using (24), a transformation equation is obtained as

$$\ddot{q}_n = -M_{22}^{-1}\left(M_{21}\ddot{\theta}_n + D_{21}\dot{\theta}_n + D_{22}\dot{q}_n + H_2 + K_2\right) \tag{25}$$

Using (25) in (23) the equation for torque is obtained as

$$A\ddot{\theta}_n + B\dot{\theta}_n + C = \tau \tag{26}$$

where the

$$\begin{cases} A = M_{11} - M_{12}M_{22}^{-1}M_{21},\ B = D_{11} - M_{12}M_{22}^{-1}D_{21}, \\ C = \left(D_{12} - M_{12}M_{22}^{-1}D_{22}\right)\dot{q}_n - M_{12}M_{22}^{-1}H_2 - M_{12}M_{22}^{-1}K_{21}\theta_n \\ \quad - M_{12}M_{22}^{-1}K_{22}q_n + H_1 + K_{11}\theta_n + K_{12}q_n \end{cases} \tag{27}$$

4.2 Design of Backstepping Based Adaptive-SMC Controller

Here is controller is designed for the tracking control of a TLFM. Considering $\theta_n = y_1$ and $\dot{\theta}_n = \dot{y}_1 = y_2$, $\dot{y}_2 = y_3$ then $y_3 = -A^{-1}(By_2 + C + \tau)$. Suppose θ_d desired trajectory for the joint angle of two-link flexible manipulator and u_v is a virtual control variable. Here desired trajectory used for (21) is a signal of chaotic system (50) given in results and discussion section. Then the trajectory tracking errors are given as

$$e_{1r} = \theta_d - y_1 \tag{28}$$

$$e_{2r} = u_v - y_2 \tag{29}$$

The derivative of the error variables given in (28) and (29) are written as:

$$\dot{e}_{1r} = \dot{\theta}_d - y_2 \tag{30}$$

$$\dot{e}_{2r} = \dot{u}_v - y_3 - A^{-1}\tau \tag{31}$$

Backstepping control input torque for the TLFM dynamics (21) is found using Theorem 1.

Theorem 1 *Consider the backstepping controller defined in* (32) *for the error dynamics* (28), (29), *the joint angle τ of* (21) *follow the desired trajectories θ_d.*

$$\tau = A(-y_3 + \dot{u}_v + c_2e_{2r} + e_{1r}) \tag{32}$$

where c_2 is a positive definite matrix.

Proof The backstepping controller for (21) is designed in below steps:
Step 1: Suppose a candidate Lyapunov function as

$$v_{1r} = \frac{1}{2}e_{1r}^2 \tag{33}$$

Using (30), the derivative of (32) is written as

$$\dot{v}_{1r} = e_{1r}\left(\dot{\theta}_d + e_{2r} - u_v\right) = e_{1r}\dot{\theta}_d - e_{1r}u_v + e_{1r}e_{2r} \tag{34}$$

Suppose the virtual control variable u is defined as

$$u_v = \dot{\theta}_d + c_1 e_{1r} \tag{35}$$

where c_1 is a positive definite matrix. We get

$$\dot{v}_{1r} = -c_1 e_{1r}^2 + e_{1r}e_{2r} \tag{36}$$

If $e_{2r} = 0$, then the first joint angle of the manipulator follows the desired trajectory.

Step 2: Considering another candidate Lyapunov function as

$$v_{2r} = v_{1r} + \frac{1}{2}e_{2r}^2 \tag{37}$$

The derivative of (37) can be written as

$$\dot{v}_{2r} = -c_1 e_{1r}^2 + e_{1r}e_{2r} + e_{2r}\left(\dot{u}_v - y_3 - A^{-1}\tau\right) \tag{38}$$

From (38), actual torque input is obtained as

$$\tau = A(-y_3 + \dot{u}_v + e_{1r} + c_2 e_{2r}) \tag{39}$$

Using (39) the derivative of (38) is writes as

$$\dot{v}_{2r} = -\left(c_1 e_{1r}^2 - c_2 e_{2r}^2\right) \tag{40}$$

Since $\dot{v}_{2r}$ in (40) is negative definite with requirement $c_1, c_2 > 0$, then the errors e_{1r} and e_{2r} asymptotically converges to origin. Thus, the joint angles of (21) follow the desired trajectories.

4.3 Design of Backstepping Adaptive SMC

In every steps of backstepping controller the system is stabilise to the desired equilibrium position by selecting suitable virtual control function and candidate Lyapunov function. At the final stage the complete system is stabilise by designing suitable control law.

In this section, a backstepping adaptive sliding mode controller is designed for faster tracking of desired trajectory and quick suppression of tip deflection.

As discussed in the problem statement section, backstepping adaptive SMC is designed at the final stage of the backstepping controller. This is designed using the error variables (30), (31) with the help of a sliding surface given in (41).

$$\sigma_b = ke_{1r} + e_{2r} \tag{41}$$

where k is positive definite matrix and is selected such that it guarantees a stable sliding surface. Stability of the sliding surface is shown by considering another Lyapunov function υ_{3r} using error variable e_1 and sliding surface σ as.

$$\upsilon_{3r} = \frac{1}{2}e_{1r}^2 + \frac{1}{2}\sigma_b^2 \tag{42}$$

Using the derivative of error (30) and sliding surface (41) in $\dot{\upsilon}_{3r}$ it is written as

$$\begin{cases} \dot{\upsilon}_{3r} &= e_{1r}\dot{e}_{1r} + \sigma_b\dot{\sigma}_b \\ &= e_{1r}(-k_{1r}e_{1r} + e_{2r}) + \sigma_b(k\dot{e}_{1r} + \dot{e}_{2r}) \\ &= e_{1r}(-k_{1r}e_{1r} + e_{2r}) + \sigma_b\left(k\dot{e}_{1r} + \dot{u}_\upsilon - y_3 - A^{-1}\tau\right) \end{cases} \tag{43}$$

Now consider the control input τ as in (44),

$$\tau = A\left(k\dot{e}_{1r} + \dot{u}_\upsilon - y_3 + \sigma_b^{-1}e_{1r}e_{2r} + \emptyset\, sign(\sigma_b)\right) \tag{44}$$

Then $\dot{\upsilon}_{3p}$ is negative definite function as given in (45).

$$\dot{\upsilon}_{3r} \leq -k_{1r}e_{1r}^2 - \emptyset|\sigma_b| \tag{45}$$

Thus, using the Lyapunov stability theorem it is seen from (45) that $\dot{\upsilon}_{3r}$ is negative definite. Hence, we can say that desired trajecking for TLFM (21) is achieved.

Now suppose the gain of the corrective control law $\emptyset$ in (44) is unknown and is estimated with time, then the control input in (44) is modified as

$$\tau = A\left(k\dot{e}_{1r} + \dot{u}_\upsilon - y_3 + \sigma_b^{-1}e_{1r}e_{2r} + \hat{\emptyset}\, sign(\sigma_b)\right) \tag{46}$$

where $\hat{\emptyset}$ is the estimate of φ. Adaptive mechanism for $\hat{\varphi}$ is designed as [38].

$$\dot{\hat{\emptyset}} = \dot{\tilde{\emptyset}} = R^{-1}(|\sigma_b|) \tag{47}$$

where $R > 0$ and $\tilde{\emptyset} = \hat{\emptyset} - \emptyset$. Now to show the stability of motion on the sliding surface with the new control law (46) and adaptive mechanism (47) is shown using another candidate Lyapunov function as in (48).

$$\upsilon_{4r} = \frac{1}{2}e_{1r}^2 + \frac{1}{2}\sigma_b^2 + \frac{1}{2}R\tilde{\emptyset}^2 \tag{48}$$

Taking the derivative of (48) and using (30), (41) and (47), it is written as

$$\begin{cases} \dot{\upsilon}_{4r} = e_{1r}\dot{e}_{1r} + \sigma_b\,\dot{\sigma_b} + R\tilde{\emptyset}\dot{\tilde{\emptyset}} \\ \dot{\upsilon}_{4r} \le -ke_{1r}^2 - \hat{\emptyset}|\sigma_b| + R^{-1}\tilde{\emptyset}(|\sigma_b|) \\ \quad\ \le -ke_{1r}^2 - \emptyset|\sigma_b| \end{cases} \tag{49}$$

Since $\dot{\upsilon}_{4r} < 0$ for $\emptyset > 0$, hence the sliding motion is stable. Therefore joint trajectory of the TLFM asymptotically tracks the desired path.

Note 1: Here adaptation mechanism is designed to estimate the gain of the switching control law of the backstepping sliding mode control law. The design of backstepping adaptive SMC for a two-link flexible manipulator presented here is attempted first in the literature.

5 Results and Discussions

In this section, results of the chaotic trajectory tracking for TLFM given in (21) are shown. The results are shown using backstepping adaptive sliding mode control and backstepping controller. Comparison of both the controller confirms that the backstepping adaptive SMC perform better than the backstepping controller in terms of smaller tracking time, lesser tip deflection and low control efforts. The parameters of a physical TLFM used for simulating the (21) are given in Table 1.

The desired trajectory used for two-link flexible manipulator (21) are signals of states of chaotic system given in (50) [42].

Table 1 Parameters used for two-link flexible manipulator (21) [41]

Link-1 mass, $m_1 = 0.15268$ Kg	Coefficients of viscous damping, $B_{eq1} = 4$Nms/rad, $B_{eq2} = 1.5$ Nms/rad
Link-2 mass, $m_2 = 0.0535$ Kg	Efficiency of gear boxes, $\eta_{g1} = 0.85$, $\eta_{g2} = 0.9$
Link-1 length, $L_1 = 0.201$ m	Efficiency of motors, $\eta_{m1} = 0.85$, $\eta_{m2} = 0.85$
Link-1 length, $L_2 = 0.201$ m	Constants of back emf, $K_{m1} = 0.119$ v/rad, $K_{m2} = 0.0234$ v/rad
Resistance of Armatures, $R_{m1} = 11.5\,\Omega$, $R_{m2} = 2.32\,\Omega$	Gear ratio, $K_{g1} = 100$, $K_{g2} = 50$
Equivalent MI at load, $J_{eq1} = 0.17043$ Kgm^2	Motor torque constants $K_{t1} = 0.119$ Nm/A, $K_{t2} = 0.0234$ Nm/A
Equivalent MI at load, $J_{eq2} = 0.0064387$ Kgm^2	Stiffness of the links, $K_{s1} = 22$ Nm/rad, $K_{s2} = 2.5$ Nm/rad
Link-1 MI, $J_{arm1} = 0.002035$ Kgm^2	Link-2 MI, $J_{arm2} = 0.0007204$ Kgm^2

$$\begin{cases} \dot{x} = y \\ \dot{y} = z \\ \dot{z} = w \\ \dot{w} = -z - aw - bwz^2 - f \end{cases} \tag{50}$$

The system in (50) is chaotic with $a = 0.5, b = 0.1$ and $x(0) = (0.02,\ 0.03,\ 0.04,\ 0.05)^T$. Chaotic attractors and time response of system (50) showing chaotic behaviour are shown in Fig. 2. Here the state x is considered as the desired signal $\theta_d = x$ in (28) for TLFM (21).

In the paper, simulations are done by using $ode-45$ solver in MATLAB simulation environment. The dynamics of TLFM (21) is simulated with the initial condition given in (49). The initial conditions used for the adaptation law is also given in (51).

$$\begin{aligned} &q(0) = (0.1, 0.2, 0, 0.0001, 0, 0.0001)^T, \\ &\dot{q}(0) = (0, 0, 0, 0, 0, 0)^T,\ \tilde{\emptyset}(0) = \hat{\emptyset}(0) = (0, 0)^T \end{aligned} \tag{51}$$

The values of the some other constants used for simulating backstepping adaptive SMC controller are

$$k_{1r} = \begin{bmatrix} 2 & 0 \\ 0 & 40 \end{bmatrix}, k = \begin{bmatrix} 2 & 0 \\ 0 & 2 \end{bmatrix}, R = (1, 1)^T \tag{52}$$

Joint trajectories tracking of the chaotic signal for both the links of the TLFM (21) are presented in Fig. 3. It is noted from Fig. 3 that the tracking of the chaotic

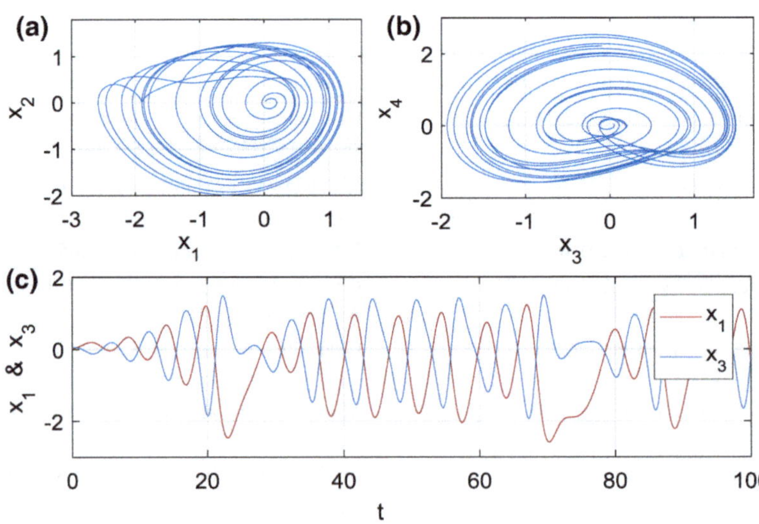

Fig. 2 Response of the system given in (50)

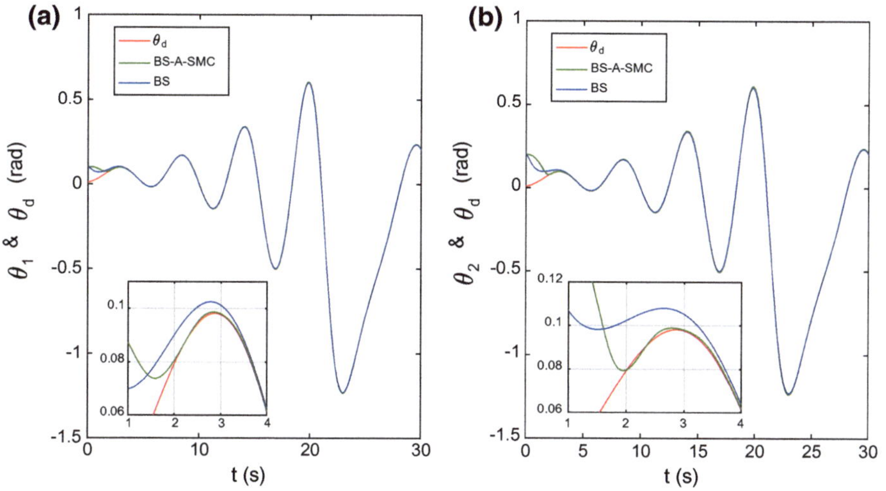

Fig. 3 Responses of chaotic trajectory tracking for the TLFM (21) of: **a** link-1 and **b** link-2 using backstepping (BS) and backstepping adaptive SMC (BS-A-SMC)

signal desired trajectory using the adaptive backstepping sliding mode control is achieved within 2 s for both the links, whereas with backstepping control the tracking is achieved within 4 s. The modes shapes of the links of the TLFM (21) are shown in Figs. 4 and 5, respectively. It is noted from Figs. 4 and 5 that deflection on modes using the backstepping adaptive SMC is less in comparison using the backstepping controller. The nature of tip deflections of the TLFM in the presence of both the backstepping adaptive-SMC and backstepping controller are given in Fig. 6. Figure 6 says that the tip deflection using the backstepping adaptive-SMC is lesser in comparison using the backstepping controller. It is also noted from Fig. 6 that the nature of deflection in Fig. 6 using backsteping adaptive SMC is less as compared with the backstepping controller.

The nature of the torque profile for the TLFM (21) used for chaotic trajectory tracking using both the backstepping adaptive MSC and backstepping controller is presented in Fig. 7. Figure 7 explains that the control input required using backstepping adaptive SMC for TLFM in the links are in the range of $[-0.4,\ 0.4]$ Nm, smoother and lesser in comparison with backstepping control. It is also apparent from Fig. 7 that the fluctuation in control torque profile using backstepping adaptive-SMC is less in comparison with backstepping control. It is noted Figs. 3, 4, 5, 6 and 7 that the backstepping adaptive SMC perform better in comparison with the backstepping controller.

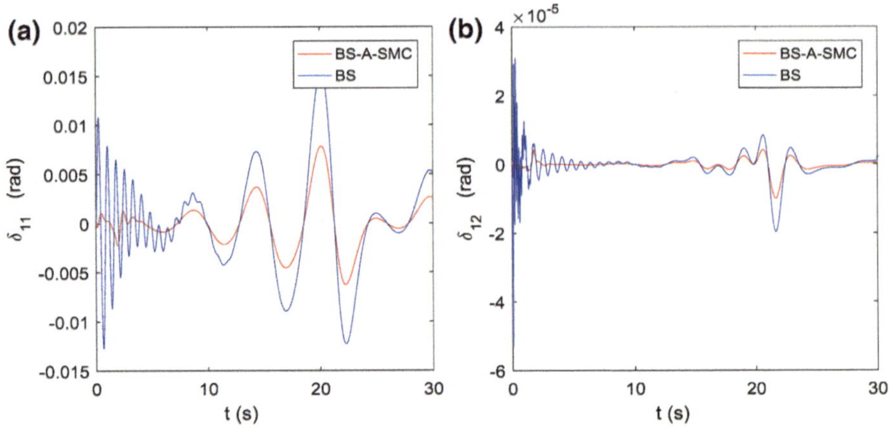

Fig. 4 Modes of the link-1 of (21) during chaotic path following using backstepping (BS) and backstepping adaptive SMC (BS-A-SMC). Desired value of the modes deflection is zero

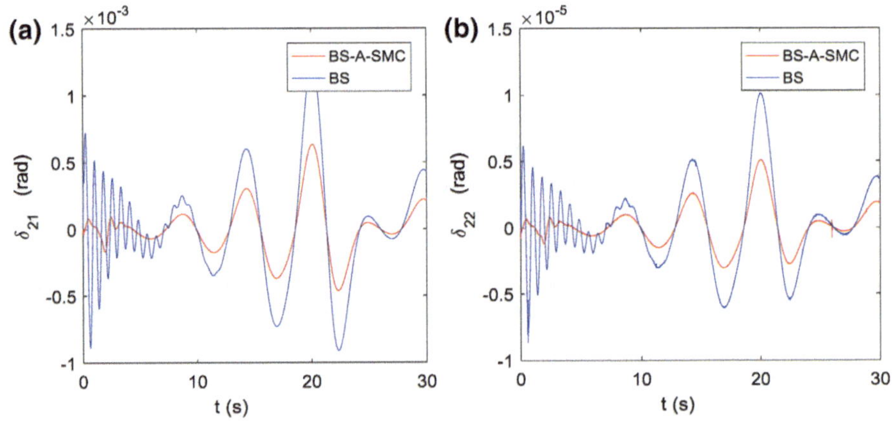

Fig. 5 Modes of link-2 of TLFM (21) when chaotic path following using backstepping (BS) and backstepping adaptive SMC (BS-A-SMC). Desired value of the modes deflection is zero

6 Conclusions

In the paper, a backstepping based adaptive-SMC is designed for chaotic signal desired tracking of a planner TLFM. Performance of the proposed backstepping adaptive-SMC is compared with an available backstepping controller. It is observed from MATLAB results that the backstepping based adaptive-SMC perform better than the backstepping controller in terms of smaller tracking time, lower link deflec-

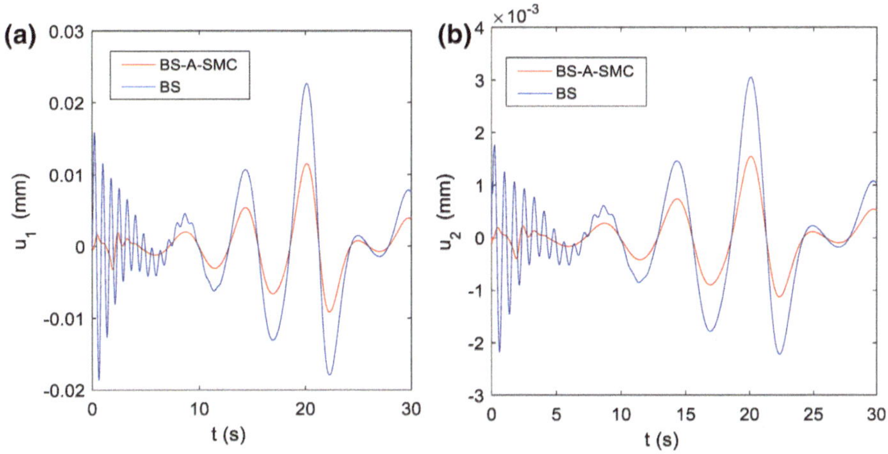

Fig. 6 Tip deflection when chaotic path following of links using backstepping (BS) controller and backstepping adaptive SMC (BS-A-SMC). Desired value of the deflection is zero

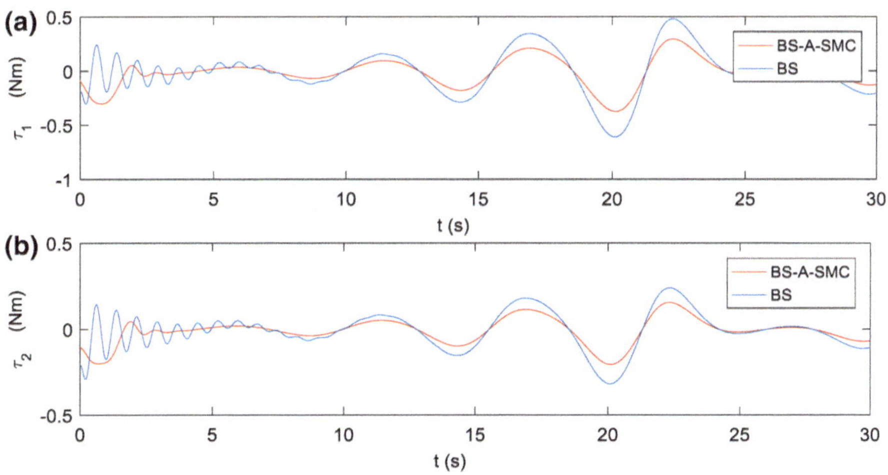

Fig. 7 Required control torque inputs using backstepping (BS) controller and backstepping adaptive SMC (BS-A-SMC) for chaotic tracking control for links of the two-link flexible manipulator (21)

tion and lesser, smother control inputs. The nonlinear dynamic model of the TLFM is derived based on the assumed modes modelling method (AMM) with assumption of two modes shape for each link.

References

1. Pradhan SK, Subudhi B (2012) IEEE Trans Autom Sci Eng 9(2):237–249
2. Kiang CT, Spowage A, Yoong CK (2014) J Intell Robot Syst Theory Appl 77(1):187–213
3. Zhang L, Liu J (2012) IET Control Theory Appl 6(13):2120–2133
4. Pradhan SK, Subudhi B (2014) IEEE Trans Control Syst Technol 22(5):1754–1768
5. Pradhan SK, Subudhi B (2014) IEEE Trans Control Syst Technol 22:1754–1768
6. Subudhi B, Morris AS (2009) Appl Soft Comput 9(1):149–158
7. Lochan K, Roy BK, Subudhi B (2016) Annu Rev Control 42:346–367
8. Bai M, Zhou DH, Schwarz H (1998) IEEE Trans Robot Autom 14(6):940–950
9. Sayahkarajy M, Mohamed Z, Faudzi AAM, Supriyanto E (2016) Eng Comput 33(2)
10. Lochan K, Suklabaidya S, Roy BK (2015) In: Proceeding of 2nd advances in robotics, vol 1, pp 2–7
11. Lochan K, Roy BK, Subudhi B (2016) IFAC-PapersOnLine 49(1):219–224
12. Lochan K, Roy BK (2015) Int J Control Theory Appl 8(3):1137–1146
13. Yu Y, Yuan Y, Fan X, Yang H (2015) Adv Sp Res 56:2312–2322
14. Husain A, Mohamed BSZ, Khairudin M (2013) J Vib Control. https://doi.org/10.1177/0733464811418094
15. Lochan K, Roy BK, Subudhi B (2017) Rob Auton Syst 97:108–124
16. Lochan K, Suklabaidya S, Roy BK (2015) In: 2nd conference on advancement in robotics, 02–04, July 2015
17. Roopaei M, Jahromi MZ (2009) Nonlinear Anal Theory Methods Appl 71(10):4430–4437
18. Van Pham C, Wang YN (2014) J Intell Robot Syst 79(1):101–114
19. Lochan K, Roy BK (2017) Int J Dyn Control, 1–25. https://doi.org/10.1007/s40435-017-0371-1
20. Lochan K, Singh JP, Roy BK, Subudhi B (2018) Nonlinear Dyn 93(4):2071–2088
21. Khalil HK (2002) Nonlinear systems, 3rd edn. Pearson Education, East Lansing
22. Soltanpour MR, Khalilpour J, Soltani M (2012) Int J Innov Comput Inf Control 8(8):5487–5498
23. Jamil MU, Noor MN, Raza MQ, Rizvi S (2014) In: 17th IEEE international multi-topic conference on collaborative and sustainable development of technologies, IEEE INMIC 2014—Proceedings, vol 1, pp 389–394
24. Su C, Stepanenko Y, Dost S (1996) IEEEI/ASME Trans Mechatron 1(4)
25. Nikdel N, Badamchizadeh MA (2017) Int J Mater Mech Manuf 5(2):113–115
26. Hu Q, Xu L, Zhang A (2012) J Franklin Inst 349(3):1087–1105
27. Soltanpour MR, Shafiei SE (2009) Elektron. ir. Elektrotechnika 8(8):75–80
28. Soltanpour MRR, Fateh MMM (2009) World Appl Sci J 6(1):70–76
29. Mbede JB, Ahanda JJM (2014) J Robot, 2014
30. Jong HO, Leet JS (1998) In: Proceedings of the 37th IEEE conference on decision & control, 1998, December, pp 931–936
31. Chang W, Tong S, Li Y (2016) Neural Comput Appl. https://doi.org/10.1007/s00521-016-2425-2
32. Sahab AR, Modabbernia MR (2011) Int J Innov Comput Inf Control 7(7):4161–4170
33. Sahab AR, Pastaki AG (2011) Int J Comput Electr Autom Control Inf Eng 5(7):791–795
34. Huang JW, Lin JS (2008) IFAC Proc 17(1):11775–11780
35. Li Y, Tong S, Li T (2013) In: IEEE international conference on control, decision and information technologies (CoDIT), pp 864–871
36. Koshkouei AJ, Burnham KJ (2016) Int J Syst Sci 42(2):1935–1946
37. Lochan K, Roy BK (2016) Int J Control Theory Appl 9(39):239–246
38. Mobayen S (2015) Nonlinear Dyn 82(1–2):599–610
39. Subudhi B, Morris AS (2002) Rob Auton Syst 41(4):257–270
40. De Luca A, Siciliano B (1991) IEEE Trans Syst Man Cybern 21(4):826–839
41. QUANSER (2006) Equation for the first (second) stage of the 2DOF serial flexible link robot. QUANSER
42. Chlouverakis KE, Sprott JC (2006) Chaos. Solitons Fractals 28(3):739–746

Multi-robot Coverage Using Voronoi Partitioning Based on Geodesic Distance

Vishnu G. Nair and K. R. Guruprasad

Abstract In this paper we propose Geodesic-VPC, a "partition" and "cover" strategy for a multi-robot system using Voronoi partitioning based on geodesic distance metric in the place of the usual Euclidean distance. Each robot is responsible for covering the corresponding geodesic-Voronoi cell using a single-robot coverage strategy. The proposed partitioning scheme ensures that Voronoi cells are contiguous even in the presence of obstacles. We demonstrate that if the single-robot coverage strategy is capable of providing a complete and non-repetitive coverage, then the proposed Geodesic-VPC strategy provides a complete and non-repetitive coverage. We use spanning tree-based coverage algorithm as the underlying single-robot coverage strategy for the purpose of demonstration, though any existing single-robot coverage algorithm can be used.

1 Introduction

Coverage path planning (CPP) for a mobile robot involves planning a path that makes the robot move through every point in the work space, and has many applications in domestic, agricultural, and defense sectors. A survey on single-robot coverage path planning algorithms is provided in [1]. Multi-robotic systems (MRS) are known for their robustness to failure of a few of the individual robots, apart from reduction in time to complete an assigned task. Several multi-robotic coverage (MRC) algorithms have been presented in the literature. An elaborate review on CPP algorithms is,

V. G. Nair
Department of Aeronautical and Automobile Engineering, Manipal Institute of Technology, Manipal Academy of Higher Education, Manipal 576104, Karnataka, India

K. R. Guruprasad (✉)
Department of Mechanical Engineering, National Institute of Technology, Surathkal, Karnataka, India
e-mail: krgprao@gmail.com

C. Shreesha and R. D. Gudi (eds.), *Control Instrumentation Systems*, Lecture Notes in Electrical Engineering 581,
https://doi.org/10.1007/978-981-13-9419-5_5

including the multi-robotic scenario, is presented in [2]. One of the main issues in a MRC is of coordination between the individual robots to ensure complete and non-repetitive coverage.

One of the approaches in MRC is "partition and cover", where the region to be covered is decomposed into cells and each robot is required to cover a cell or a set of contiguous cells. Major advantages of the "partition" and "cover" strategy are ease of task allocation and hence, reduced or no requirement of inter-robot communication during the coverage process, and reduced memory requirement due to task partitioning. The problem of coordination between the robot is trivial in these approaches. The main advantages of using Voronoi partitioning for coverage, as in [3], are that the robot is guaranteed to be within the cell it has to cover unlike in the other approaches where the robot may initially lie outside the cell/region allotted for it to cover, and the area to be covered may be partitioned in a distributed manner. However, one of the main disadvantages of this approach is non-uniform cell size. This issue can be addressed by ensuring the robots are reasonably uniformly distributed over the area of interest before partitioning. Another major disadvantage of using Voronoi partition is that the partitioning schemes do not take into account the obstacles in the region. Presence of obstacles may create situations where a cell may contain several topologically disconnected patches (that is, non-contiguous). Several approaches have been proposed in the literature to ensure that each robot is assigned a task of covering a contiguous (or topological connected) region. In [4], a distributed repartitioning scheme is proposed based on the principle of the standard Voronoi partitioning scheme. In [5], the problem of topologically disconnected Voronoi cells is handled using a dynamically repartitioning the region using an auction protocol.

In this paper we propose a multi-robot coverage algorithm based on the "partition" and "cover" approach using geodesic distance based Voronoi partition to alleviate the problem of disconnected cells in the presence of obstacles. This eliminates the need for repartitioning.

2 Voronoi Partitioning and Coverage Problem

Voronoi partitioning is a scheme of area partitioning widely used in several problems such as facility location/locational optimization, robot path planning, multi-agent/robotic systems, sensor networks, etc. A partition of set X is a collection of subsets W_i with disjoint interiors such that their union is X itself.

Let $I_N = \{1, 2, \ldots, N\}$; $Q \subset \mathbb{R}^2$; and $\mathcal{P} = \{p_1, p_2, \ldots, p_N\}$, $p_i \in Q$, be a set of points in Q called a *node set*. The *Voronoi partition*, generated by $\mathcal{P}$ is the collection $\{V_i(\mathcal{P})\}_{i \in I_N}$ with,

$$V_i(\mathcal{P}) = \left\{ q \in Q | \parallel q - p_i \parallel \leq \parallel q - p_j \parallel, \forall j \in I_N \right\} \tag{1}$$

The collection of all the points closest to the node p_i is the Voronoi cell corresponding to the node p_i. One of the main properties of Voronoi cells which is that each Voronoi cell is a topologically connected non-null set. In the context of multi-robot coverage

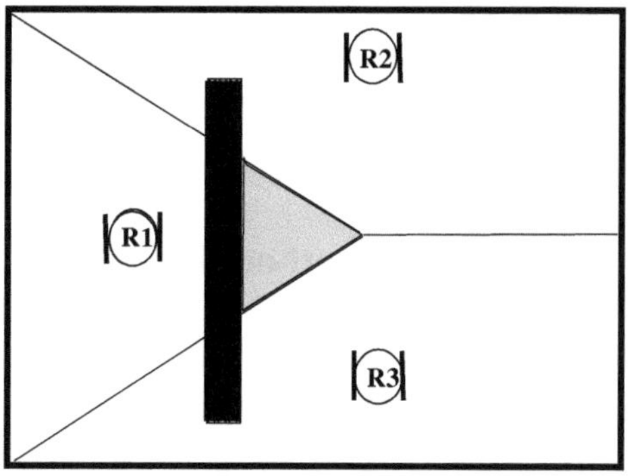

Fig. 1 The standard Voronoi partitioning with 3 robots (position of robots are used as nodes) in a workspace occupied by an obstacle. The Voronoi cell corresponding to the 'robot R_1' is made up two disconnected patches separated by the obstacle

problem addressed here, initial positions of robots are used as nodes, and each robot is assigned the task of covering the corresponding Voronoi cell [3]. The fact that the Voronoi cells partition an area ensures that the coverage is complete and non-repetitive if each robot covers the corresponding Voronoi cell completely without any coverage overlap.

Note that the partitioning scheme does not distinguish between free space and space occupied by the obstacles. The entire space, which may include obstacles (known a priori or not), is partitioned. Thus, a Voronoi cell may contain obstacles within it. An obstacle may split a Voronoi cell into two or more topologically disconnected patches of free space as illustrated in Fig. 1. Here, the region of interest is partitioned into Voronoi cells based on nodes $R1$, $R2$, and $R3$ (which are positions of three robots in this situation). The presence of an obstacle splits the V_1, the Voronoi cell corresponding to node/robot $R1$ into two topologically disconnected patches. If the robot $R1$ has to reach a point shaded grey, a portion of its own Voronoi cell, it has to pass through V_2 or V_3. In the context of a multi-robot coverage problem addressed in this paper, $R1$ crossing its Voronoi boundary and moving over the neighboring Voronoi cell, which is being covered by $R2$ (or $R3$), leads to coverage overlap. Further, when robot $R1$ moves in V_2 (or V_3), it has to ensure that it does not collide with the robot $R2$ (or $R3$), unlike in a situation where each robot has to move (cover) only within the corresponding Voronoi cell.

Such situations lead to sub-optimal solutions to the coverage problem, and hence increases the time and energy required to complete the task. To avoid coverage overlap, we need to ensure that each robot is assigned a contiguous region to cover. Repartitioning as in [4, 5] is one possible solution. A more elegant and easier solution is to incorporate the knowledge of the obstacles into partitioning scheme and ensuring that each cell has a single topologically connected patch of free space. In other words, instead of partitioning the entire region Q, partition only the free space $Q \setminus O$, where O represents the region occupied by obstacles.

Geodesic distance based generalization of Voronoi partition has been used in many applications such as in sensor coverage and sensor placement problems [6–11]. In [12] author discusses algorithm for computing geodesic distance based Voronoi partitions. In this paper, we propose to use geodesic distance based Voronoi partitioning to overcome the problem of topological disconnected Voronoi cells associated with the standard Euclidean distance.

3 Proposed Methodology

In this section, we present the proposed multi-robot coverage strategy based on geodesic Voronoi partition. In geodesic distance based generalization of Voronoi partitioning scheme, geodesic distance between points is considered in the place of the Euclidean distance.

The term *geodesic* in its original form comes from *geodesy*, which is the science of measuring the size and shape of Earth. A geodesic in this sense is the shortest route between two points on the Earth's surface. Unlike on a flat surface, on the Earth's surface, shortest route between two points is not a straight line. This generalization of shortest distance between two points (length of the straight line segment on a flat surface) is known as the *geodesic distance*. In general, geodesic distance between any two points is the length of the shortest path between them. In the context of a mobile robot moving on a flat surface containing obstacles, the concept of geodesic distance is still useful. Though the Euclidian distance is still valid on the region of interest, due to the presence of obstacles, a robot may not be able to move between two points along a straight line. In this scenario, the geodesic distance is defined as the shortest path between two points in question that avoids the obstacles. Actual path and hence the geodesic distance depends on the specific path planning algorithm the robot uses. In this paper, for simplicity, we assume that the obstacles are polygonal in shape, and hence the shortest path between any two points is always a sequence of line segments. Such scenarios are commonly used in the literature in similar situations [10]. The geodesic distance used in this paper is illustrated in Fig. 2.

Consider $Q \subset \mathbb{R}^2$, a compact (that is, closed and bounded), not necessarily a convex set, as the region of interest. Let $O = \bigcup_{i=1}^{m} O_i$, be the region within Q occupied by the m (polygonal) obstacles. Let $d_G(q_1, q_2)$ denote the geodesic distance between points $q_1, q_2 \in Q \setminus O$. Further, let $\mathcal{P} = \{p_1, p_2, \ldots, p_N\}$, $p_i \in Q \setminus O$, be the node set. Now a geodesic distance based Voronoi partition of $Q \setminus O$, the free space within Q is given by $\{V_1^G, V_2^G, \ldots, V_n^G\}$. Here,

$$V_i^G(\mathcal{P}) = \left\{q \in Q \setminus O | d_G(q, p_i) \le d_G(q, p_j), \forall j \in I_N\right\} \tag{2}$$

Observe that, unlike in the case of the standard Voronoi partition which uses the Euclidean distance metric, geodesic distance based Voronoi partition scheme decomposes the free space $Q \setminus O$ rather that the whole of the region Q. Though the standard Voronoi cell is a topologically connected region within Q, as observed in the previous section, it may lead to non-connected region in $Q \setminus O$. Now as the geodesic Voronoi partition scheme decomposes only the obstacle-free space within Q, the corresponding cells are always topologically connected. This is illustrated

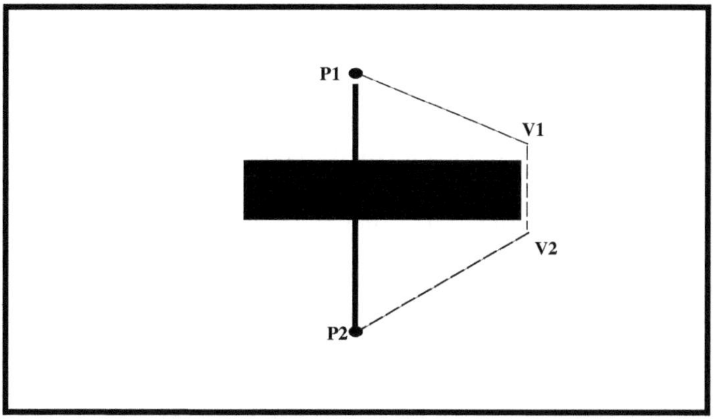

Fig. 2 Euclidean distance between two points P_1 and P_2 is the straight line joining them, while the geodesic distance is the shortest obstacle-free path from P_1 to P_2, which is made of line segments $P_1 --> v_1 --> v_2 --> P_2$

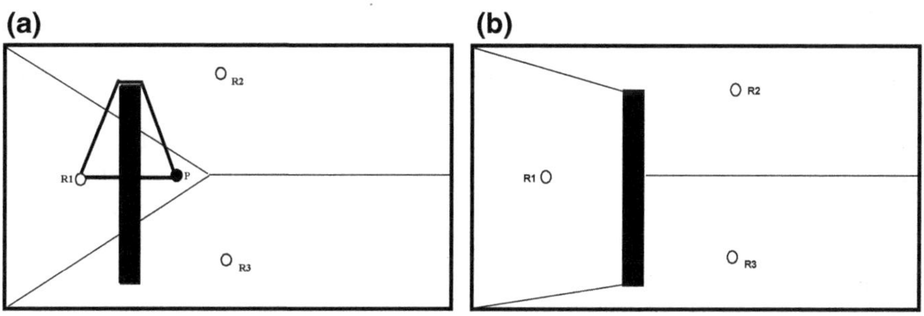

Fig. 3 Standard Voronoi and Geodesic Voronoi illustration, **a** standard Voronoi partition may lead to non-contiguous Voronoi cells. Though the point P is close to R_1 in the Euclidian sense, actual robot path is that avoiding the obstacle (that is, the geodesic path) and covering a larger distance. Though the point P lies in V_1 the Voronoi cell corresponding to R_1, the robot has to pass through V_2 (or V_3), to reach this point and **b** the partitioning based on geodesic distance ensures that each Voronoi cell is a contiguous region

using an example in Fig. 3. The standard Voronoi partition shown in Fig. 3a, partitions entire region (Q is a rectangular region here), and the Voronoi cell corresponding to node 1 (Shown as $R1$, the 'robot 1') within the obstacle-free region is split into two topologically disconnected patches by the presence of an obstacle. The same region when decomposed using the geodesic distance based Voronoi partition results in topologically connected cells, as shown in Fig. 3b.

Once the geodesic distance based Voronoi cells are computed, each robot uses a suitable single-robot coverage algorithm to cover the respective Voronoi cell. As in any "partition" and "cover" algorithm, once the partitioning is done, the problem is reduced to a single-robot coverage problem.

In addition to reduced time for completion of the given task (cover the region Q here), multi-robotic systems have another very useful property of being robust to failure of a few individual robots. That is, the given task (cover the region Q) may be completed even if a few of the individual robots fail. We may use technique used in [3] to incorporate this property to the proposed Geodesic-VPC. Each robot broadcasts 'I am alive' at regular intervals. If a robot fails, it will naturally stop sending such as message. Now the neighboring robots may repartition the region and complete the coverage. In this paper, we do not focus on the robustness property.

Note that to compute the geodesic distance based Voronoi partition, the information about the obstacles should be available a priori. This information may be either known or may be gathered by robots by performing an exploration. In this paper, we assume that the map of the region (that is, exact location of obstacles) is known. We are currently working on the scenarios when the robot may start with no knowledge of the map and build the map while covering the region, and updating the Voronoi cells as and when the new information about the obstacles is obtained.

Another aspect of the multi-robot coverage problem is of uniform load distribution amongst the individual robots in terms of coverage time. This problem is common to any multi-robot coverage algorithms, particularly the "partition" and "cover" class of algorithms. The problem requires uniform partitioning of the region, known as equatable partitioning. More uniform coverage can be achieved either by placing the robots more uniformly in the region before starting the coverage, or uniformly placed virtual nodes in the region and partitioning the region based on these virtual nodes. Several strategies such as centroidal Voronoi configuration have been explored in the literature to achieve uniform area partitioning. In this paper, we do not consider the problem of deployment or placement of robots.

4 Results and Discussion

In this section we provide illustrative examples to demonstrate the proposed geodesic-VPC strategy using spanning tree-based coverage (STC) [13] algorithm as the underlying single-robot coverage strategy. This algorithm is used here only for the purpose of illustration/demonstration. Any single-robot coverage algorithm may be used for the proposed geodesic-VPC. In fact the processes of *partitioning* and *coverage* are decoupled. The overall coverage performance (in terms of completeness and overlap) depends on the single-robot coverage algorithm. However, Geodesic-VPC strategy ensures that there is no duplication of coverage between any two robots and also each robot needs to cover a contiguous region. We have used Matlab for geodesic Voronoi partitioning and path planning using the single-robot coverage algorithm. The path generated by the path-planning algorithm over the graph formed by the grids is shown rather than actual path followed by the robot.

Figure 4 shows coverage path generated using STC algorithm again in two scenarios, (a) with a line obstacle, and (b) with a triangular obstacle. In STC, we divide the region of interest into square grids called minor cells of size $D \times D$, which is the

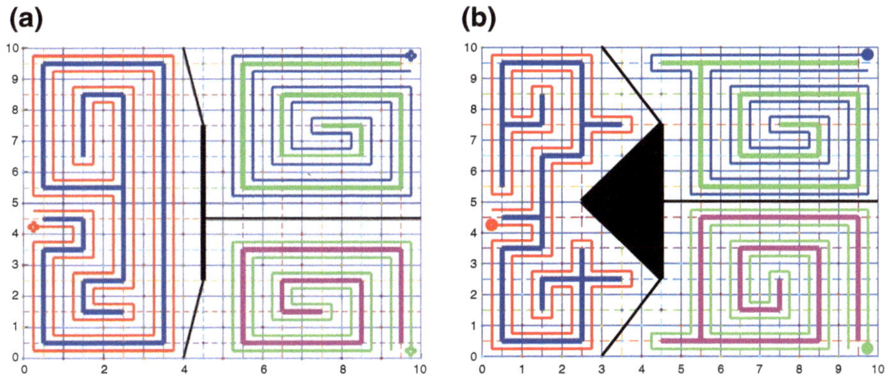

Fig. 4 Coverage path generated by STC algorithm using geodesic Voronoi partitioning of the workspace in presence of obstacle using three robots in two scenarios, **a** with a line obstacle, and **b** with a triangular obstacle. The robot path is shown by thinner lines around the spanning tree created (Shown with thick line) within each of the Geodesic distance based Voronoi cells). STC provides non-overlapping coverage within each Geodesic Voronoi cell, with coverage gaps near the partition/obstacle boundaries

size of the coverage tool footprint. Four minor cells are combined to form a major cell/grid of size $2D \times 2D$. In each of the geodesic Voronoi cells, each of completely free major grid is covered and there are no repetitive paths. However, the STC algorithm does not create path through partially occupies major grids, and leading to incomplete coverage near the partition/obstacle boundaries.

We may observe that the geodesic Voronoi partitioning provides a contiguous region for coverage by individual robots and also, there is no overlap between the coverage area of any two robots. A single-robot coverage algorithms such as that given in [14], which provides a complete coverage of each cell without unnecessary coverage overlap.

5 Conclusions

We proposed Geodesic-VPC, a "partition" and "cover" multi-robot area coverage strategy, using geodesic distance based Voronoi partitioning scheme, in the presence of obstacles. Each robot is allotted the task of covering of a Geodesic Voronoi cell. Unlike the standard Voronoi cell (based on the Euclidean distance), the geodesic Voronoi cell is a contiguous region in the free space. As each robot covers the corresponding geodesic Voronoi cell, a passive cooperation between the robots is achieved, thus avoiding coverage duplication and without any requirement of extensive communication during the coverage process. Also, as each robot has to cater to a smaller region and does not require the information of the coverage map of other robots, the memory requirement is also greatly reduced.

References

1. Choset H (2001) Coverage for robotics—a survey of recent results. Ann Math Artif Intell 31(1–4):113–126
2. Galceran E, Carreras M (2013) A survey on coverage path planning for robotics. Robot Auton Syst 61:1258–1276
3. Guruprasad KR, Wilson Z, Dasgupta P (2012) Complete coverage of an initially unknown environment by multiple robots using voronoi partition. In: Proceedings of 2nd international conference on advances in control and optimization of dynamical systems
4. Guruprasad KR, Dasgupta P (2012) Distributed spatial partitioning of an initially unknown region for a multi-robot coverage application. In: Autonomous robots and multirobot systems (ARMS) workshop (co-located with AAMAS 2012)
5. Hungerford K, Dasgupta P, Guruprasad KR (2014) A repartitioning algorithm to guarantee complete, non-overlapping planar coverage with multiple robots. In: Proceedings of DARS
6. Bhattacharya S, Michael N, Kumar V (2013) Distributed coverage and exploration in unknown non-convex environments. In: Distributed autonomous robotic systems, pp 61–75
7. Fekete SP, Kamphans T, Kroller A, Mitchell J, Schmidt C (2011) Exploring and triangulating a region by a swarm of robots. In: Proceedings of APPROX, LNCS, vol 6845, pp 206–217
8. Breitenmoser A, Schwager M, Metzger J-C, Siegwart R, Rus D (2010) Voronoi coverage of non-convex environments with a group of networked robots. In: Proceedings of IEEE international conference on robotics and automation, pp 4982–4989
9. Lee SK, Fekete SP, McLurkin J (2014) Geodesic topological Voronoi tessellations in triangulated environments with multi-robot systems. In: Proceedings of IEEE international conference on robotics and automation
10. Pimenta LCA, Kumar V, Mesquita RC, Pereira GAS (2008) Sensing and coverage for a network of heterogeneous robots. In: 2008 47th IEEE conference on decision and control, pp 3947–3952
11. Thanou M, Stergiopoulos Y, Tzes A (1989) On the geodesic Voronoi diagram of point sites in a simple polygon. In: Algorithmica, pp 109–140
12. Aronov B (2013) Distributed coverage using geodesic metric for non-convex environments. In: 2013 IEEE international conference on robotics and automation, pp 933–938
13. Gabriely Y, Rimon E (2001) Spanning tree based coverage of continuous areas by a mobile robot. Ann Math Artif Intell 31:77–98
14. Ranjitha TD, Guruprasad KR (2015) CCSTC: a truly complete competitive spanning tree coverage algorithm for mobile robots. In: Proceedings of advances in robotics, 2nd international conference of robotics society of India. BITS Goa, India, July 2015

Secure Communication Using a New Hyperchaotic System with Hidden Attractors

Jay Prakash Singh, Kshetrimayum Lochan and Binoy Krishna Roy

Abstract Objectives of the paper are to (i) develop a new hyperchaotic system having hidden attractors and (ii) to show the applications using the new system in the form of secure communication. New system proposed in the paper has a stable equilibrium, hence considered under the class of the hidden attractors dynamical system. Dynamical characteristics of the novel system is confirmed using some numerical means like phase portrait, Poincaré map and Lyapunov spectrum plot. The applications of the new system are shown by encrypting and decrypting a sinusoidal signal and sound wave. Secure communication is achieved by designing a proportional integral (PI) based sliding mode control (SMC). MATLAB simulation results validate and ensure that the objectives are satisfied.

Keywords New hyperchaotic system · Hidden attractors · Sliding mode control · Secure communication · Control of chaos

1 Introduction

Available dynamical characteristic chaotic systems are classified into two clusters. The reported systems with (i) hidden attractors or (ii) self-excited attractors [1–8] are the two main cluster. Finding and advancement of the systems with hidden attractors is more challenging as compared to the other part. This is because in hidden

J. P. Singh
Department of Electrical Engineering, Rewa Engineering College, Rewa 486002, MP, India
e-mail: jayprakash1261@gmail.com

K. Lochan (✉)
Department of Mechatronics, Manipal Institute of Technology, Manipal 576104, Karnataka, India
e-mail: lochan.nits@gmail.com

B. K. Roy
Department of Electrical Engineering, National Institute of Technology Silchar, Silchar 788010, Assam, India
e-mail: bkr_nits@yahoo.co.in

C. Shreesha and R. D. Gudi (eds.), *Control Instrumentation Systems*,
Lecture Notes in Electrical Engineering 581,
https://doi.org/10.1007/978-981-13-9419-5_6

attractors the knowledge of location of equilibrium point does not help in creation of the attractors [1–8]. Lorenz [9], Chen [10], Lu [11], chaotic system and systems in [12–17] are the types of self-excited attractors. Systems having stable equilibrium point [18, 19] or no equilibrium point [20, 21] are the main types of hidden attractors. Newly the chaotic/hyperchaotic systems with infinite equilibria also belong to the choice of the hidden attractors [22–27]. The study of the chaotic/hyperchaotic systems having such nature is significant. This is so because in such system unexpected and undesired behaviours can be observed [6, 28, 29].

Hidden attractors are seen in various types of chaotic/hyperchaotic system as discussed in the literature [30, 31]. Dynamical systems with stable equilibrium point hidden attractors are comparatively less available in the literature as compared with no equilibrium point system of hidden attractors. We know that the hyperchaotic systems are more complex as compared with the chaotic systems [24, 32]. Thus, development of hyperchaotic system is more important. The available dynamical system (chaotic/hyperchaotic) systems having stable nature of equilibria are given in the Table 1. Table 1 reflects that hyperchaotic systems having stable nature of the equilibrium points are very few in the literature. It is noted from the available literature and the Table 1 that there is still some scope for developing the system with stable equilibria. Considering the above discussion, this paper needs to report a new hyperchaotic system. The important feature in the new system is that it has a stable equilibrium point.

The remaining paper goes like this. Section 2 presents the dynamics of the proposed new system having hyperchaotic behaviour and stable equilibria. Numerical analysis of the reported system is discussed in Sect. 3. Application of the new system is discussed in the Sect. 4 of the paper. Results and discussion of the application is presented in Sect. 5 of the paper. And in the last the paper is concluded in the Sect. 6 of the paper.

Table 1 The dynamical systems (hyperchaotic/chaotic) having stable nature of equilibria

3-D/4-D/5-D	Types of system	References
3-D	With 1 equilibrium point	[30, 31, 33–43]
	With 2 equilibrium points	[18, 35, 44–47]
4-D	Hyperchaotic system having one equilibrium point	[48, 19, 49]
4-D	Memristive hyperchaotic system with infinitely many equilibrium points	[23]
5-D	New hyperchaotic system having two equilibria	[50]

2 A New Dynamical Hyperchaotic System Having Stable Nature of Equilibrium Point

In the present section, the dynamics of a new system is presented which is considered in the work. The new proposed system is developed from the Lorenz-stenflo system [19] by using state feedback control. The dynamics of the reporting system is presented in (1).

$$\begin{cases} \dot{x}_1 = a(x_2 - x_1) + bx_2x_3 + x_3x_4 \\ \dot{x}_2 = -bx_1x_3 + cx_2 + x_4 \\ \dot{x}_3 = 4 + x_1x_2 - dx_3 \\ \dot{x}_4 = -ex_2 \end{cases} \tag{1}$$

In system (1), a, b, c, d, e are the parameters and x_1, x_2, x_3, x_4 are considered as the state. The system in (1) is obtained from the Lorenz-stenflo system using state feedback control and perturbing one term.

The system in (1) is invariant when $(x_1, x_2, x_3, x_4) \rightarrow (-x_1, -x_2, x_3, -x_4)$. Therefore, the proposed system dynamics has regularity around the x_3 axis.

The new hyperchaotic system proposed in the paper is a dissipative dynamical system. This is proved by finding the divergence of the new system and is given in (2).

$$\nabla v = \frac{\partial \dot{x}_1}{x_1} + \frac{\partial \dot{x}_2}{x_2} + \frac{\partial \dot{x}_3}{x_3} + \frac{\partial \dot{x}_4}{x_4} = -a - c - d = -(a + c + d) \tag{2}$$

It is seen from (2) that the divergence is negative because a, c, d are the positive constants. Thus the volume in the phase space of the new system decays exponentially with the rate $(a + c + d)$. Therefore it may be said that there can be attractors in system (1).

Equilibrium point of the new considered system is found out by putting the derivate of each state variable to zero. The system in (1) has only equilibrium point at $E = \left(0,\ 0,\ -\frac{d}{4},\ 0\right)$. Eigenvalues of the new system is found out using the Jacobian matrix given in (3).

$$J_1 = \begin{bmatrix} -a & a + b(x_3)^* & b(x_2)^* + (x_4)^* & (x_3)^* \\ -(x_3)^* & 17 & -(x_1)^* & 1 \\ (x_2)^* & (x_1)^* & -d & 0 \\ 0 & -e & 0 & 0 \end{bmatrix} \tag{3}$$

Table 2 presents the eigenvalues of the system given in (1). It is specious from Table 2 that the new system has all the eigenvalues with stable nature. Thus, the new system may have hidden attractors.

Table 2 Equilibrium point and eigenvalues of system (1) with $a = 35,\ b = 30,\ c = 17,\ d = 0.78,\ e = 14$

Equilibrium point	Eigenvalues
$E = \left(0,\ 0,\ -\frac{d}{4},\ 0\right)$	$\lambda_1 = -0.78$
	$\lambda_2 = -16.70367158$
	$\lambda_3 = -0.648164 + 2.719171i$
	$\lambda_4 = -0.648164 - 2.719171i$

3 Dynamical Analysis of System (1)

Dynamical behaviour of the considered new proposed system is shown in the present section using some of the numerical method.

The new system has hyperchaotic behaviour with $a = 35,\ b = 30,\ c = 17,\ d = 0.78,\ e = 14$. Finite-time LEs for these sets of parameters are $L_i = (1.014,\ 0.218,\ 0,\ -19.686)$. Hyperchaotic attractors of the new system with $a = 35,\ b = 30,\ c = 17,\ d = 0.78,\ e = 14$ are revealed in Fig. 1. Poincaré map across $x_1 = 0$ plane of the new system is presented in Fig. 2. The dynamical behaviour/characteristics of the new system is investigated by plotting the finite-time Lyapunov spectrum (LS). The finite-time LS is plotted by finding the Lyapunov exponents with the fixed initial conditions $x(0) = (0.2,\ 0.1,\ 5,\ 0.1)^T$ and observation time $T = 20{,}000$ time unit. The LEs are calculated by the method of Wolf et al. algorithm [51] in MATLAB simulation environment. The finite-time Lyapunov spectrum with varying e keeping other parameter fixed in Fig. 3. Presence of the two positive natures of the Lyapunov exponents in Fig. 3 indicates the existence of hyperchaotic behaviour in the new system.

4 Secure Communication Using the System in (1)

Here, an application using the new proposed system is illustrated. The application is shown in the field of secure communication by masking and retrieving of information signals.

In last decade chaotic/hyperchaotic systems are commonly being applied for secure communication [52–54]. The chaotic signals are being used in various ways in the secure communication. One common way is for encryption and decryption of a message signal. The main reason for this is that chaotic signals have apparently noise like nature and unpredictable behaviour [52–55].

In the paper the secure communication using the new system in (1) is presented by considering as like the system acting as a master and system acting like as a slave system. Suppose the system acting like as a master and the system acting like as a slave system are given in (4) and (5), respectively

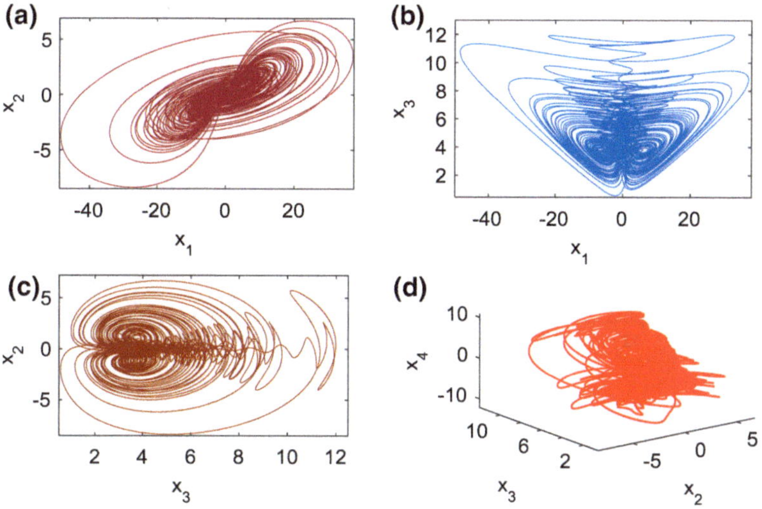

Fig. 1 Hyperchaotic attractors with $a = 35$, $b = 30$, $c = 17$, $d = 0.78$, $e = 14$ for system (1)

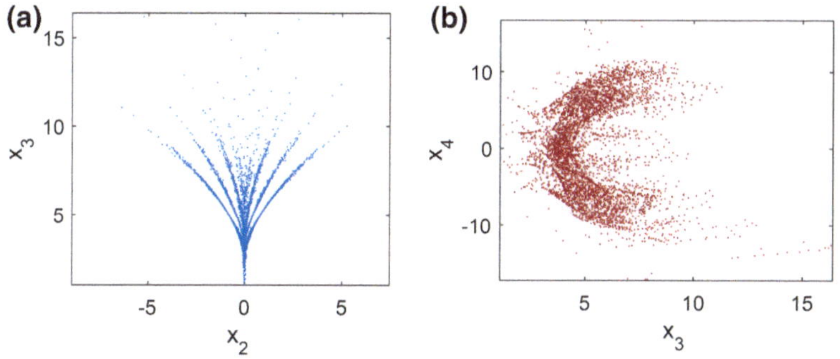

Fig. 2 Poincaré map across $x_1 = 0$ in: a $x_2 - x_3$ plane and **b** $x_3 - x_4$ plane of the new system

$$\begin{cases} \dot{x}_1 = a(x_2 - x_1) + bx_2x_3 + x_3x_4 \\ \dot{x}_2 = -bx_1x_3 + cx_2 + x_4 \\ \dot{x}_3 = 4 + x_1x_2 - dx_3 \\ \dot{x}_4 = -ex_2 \end{cases} \tag{4}$$

$$\begin{cases} \dot{y}_1 = a(y_2 - y_1) + by_2y_3 + y_3y_4 + u_1 \\ \dot{y}_2 = -bmy_3 + cy_2 + y_4 + u_2 \\ \dot{y}_3 = 4 + my_2 - dy_3 + u_3 \\ \dot{y}_4 = -ey_2 \end{cases} \tag{5}$$

where $m = x_1 + s$ and s is the message signal and u_1, u_2, u_3 control inputs which are needed to be designed. Here, the message (m) is added with state x_1. When the

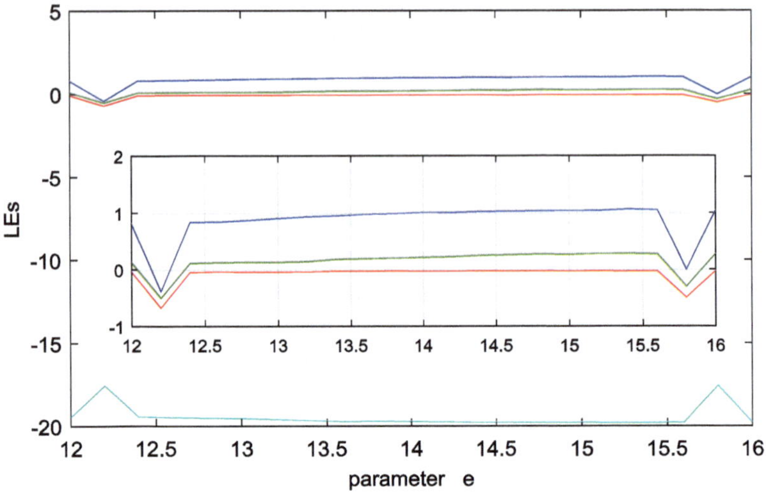

Fig. 3 Finite time LS with $a = 35,\ b = 30,\ c = 17,\ d = 0.78$ and $x(0) = (0.2, 0.1, 5, 0.1)^T$ for the new system

system acting like a master (4) has states synchronised with the system acting like the system as a slave (5) systems i.e., $y_i = x_i$, then at the receiver end the message signal $\tilde{s}$ is retrieved as $\tilde{s} = m - y \approx s$. Here it is considered that the SNR of the message signal (m) is less than the masking signal x_1. The application is completely illustrated and sketched in Fig. 4.

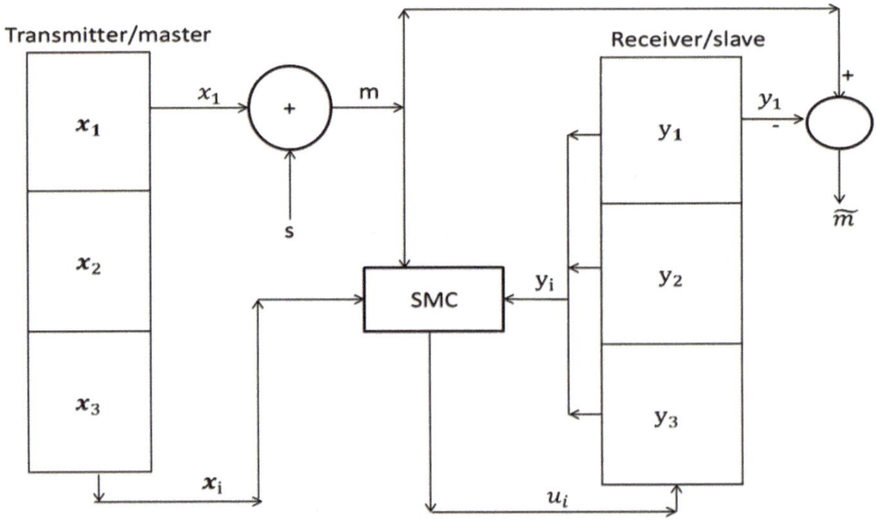

Fig. 4 Complete communication scheme [52]

The synchronisation errors among the system acting like as a master (4) and the system acting like as a slave system (5) are given in (6).

$$\begin{cases} \dot{e}_1 = a(e_2 - e_1) + be_2y_3 + bx_2e_3 + e_3y_4 + x_3e_4 + u_1 \\ \dot{e}_2 = -e_1y_3 - x_1e_3 + ce_2 + e_4 + u_2 \\ \dot{e}_3 = e_1y_2 + x_1e_2 - de_3 + u_3 \\ \dot{e}_4 = -ee_2 \end{cases} \tag{6}$$

Next question is to narrate the stabilisation of the error dynamics given in (6). It is required to bring the error dynamics to zero. The answer for the question and the required task is performed by designing a suitable SMC. Here proportional integral SMC is designed for this purpose.

The mathematical structure of the PI sliding surface is presented in (7).

$$\begin{cases} s_1 = e_1 + \int_0^t (k_1e_1)d\tau \\ s_2 = e_2 + \int_0^t (k_2e_2 - ce_4)d\tau \\ s_3 = e_3 + \int_0^t (k_3e_3)d\tau \end{cases} \tag{7}$$

where k_1 and k_2 are the user defined positive parameters. It is needed that when dynamics goes through the sliding variable, it requires to satisfies $\dot{s}_i = 0$. Now the equivalent mode dynamics [56] is be written as (8).

$$\begin{cases} \dot{e}_1 = -k_1e_1 \\ \dot{e}_2 = -(k_2e_2 - ee_4) \\ \dot{e}_3 = -(k_3e_3) \\ \dot{e}_4 = -ee_2 \end{cases} \tag{8}$$

The stabilisation of the error dynamics defined in (8) is shown by choosing a Lyapunov function candidate as $V_1(e) = \frac{1}{2}\left(e_1^2 + e_2^2 + e_3^2 + e_4^2\right)$. The $\dot{V}_1(e)$ along with (8) is written in (9).

$$\begin{cases} \dot{V}_1(e) = e_1\dot{e}_1 + e_2\dot{e}_2 + e_3\dot{e}_3 + e_4\dot{e}_4 \\ = e_1(-k_1e_1) + e_2(-(k_2e_2 - ee_4)) + e_3(-(k_3e_3)) + e_4(-ee_2) \end{cases}$$

After arranging some terms, we get

$$\dot{V}_1(x) = -k_1e_1^2 - k_2e_2^2 - k_3e_3^2 - k_4e_4^2 \tag{9}$$

where k_1, k_2, k_3, k_4 are the positive constant. It is apparent from (9) that it is negative definite. Therefore, the sliding motion is asymptotically stable.

SMC controllers proposed are designed in (10).

$$\begin{cases} u_1 = -ae_2 - be_2y_3 - bx_2e_3 - e_3y_4 - x_3e_4 - k_1e_1 - \rho_1 \tanh(\sigma_1) \\ u_2 = e_1y_3 + x_1e_3 - ce_2 - (1-e)e_4 - k_2e_2 + sy_3 - \rho_2 \tanh(\sigma_2) \\ \quad u_3 = -e_1y_2 - x_1e_2 - k_3e_3 - sy_2 - \rho_2 \tanh(\sigma_2) \end{cases} \tag{10}$$

Theorem 1 The error in (6) converges to $\sigma_i = 0$ if it is controlled by (10) and also ensure synchronisation between the system as the master (4) and the system as the slave (5) system.

Proof Suppose Lyapunov candidate function as $V_2(s) = \frac{1}{2}\left(s_1^2 + s_2^2 + s_3^2\right)$. The time derivative of $V_2(s)$ along with (7) can be written as

$$\begin{cases} \dot{V}_2(S) = S_1\dot{S}_1 + S_2\dot{S}_2 + S_3\dot{S}_3 \\ = s_1(\dot{e}_1 + k_1e_1) + s_2(\dot{e}_2 + k_2e_2 - ce_4) + s_3(\dot{e}_3 + k_3e_3) \\ = s_1(a(e_2 - e_1) + be_2y_3 + bx_2e_3 + e_3y_4 + x_3e_4 + u_1 + k_1e_1) \\ \quad + s_2(-e_1y_3 - x_1e_3 + ce_2 + e_4 + u_2 + k_2e_2 - ce_4) + \\ \quad\quad s_3(e_1y_2 + x_1e_2 - de_3 + u_3 + k_3e_3) \end{cases} \tag{11}$$

Now inserting the control laws (10) in (11) we get,

$$\begin{aligned} \dot{V}_2(s) &= -\sigma_1 s_1 \tanh(s_1) - \sigma_2 s_2 \tanh(s_2) - \sigma_3 s_3 \tanh(s_3) \\ &< -\rho_1|s_1| - \rho_2|s_2| - \rho_3|s_3| < 0 \end{aligned} \tag{12}$$

where ρ_1, ρ_2, ρ_3 are the positive constants. Thus, we can say that $\dot{V}_2(s) < 0$ for $s \neq 0$. Therefore sliding surfaces s_1, s_2 and s_3 converge to $s_1 = 0$, $s_2 = 0$ and $s_3 = 0$, [56] respectively. Hence, error dynamics given in (7) stabilises at origin. Therefore, the master (4) and the slave (5) systems states are synchronised. Therefore the message signal is encrypted and decrypted successfully using the proposed approach.

5 Results and Discussion for Secure Communication Using the New Hyperchaotic System

This section discussed the secure communication using the new hyperchaotic system. The application is shown by encryption and decryption of a sinusoidal signal and sound like signal. Simulation of the master and slave hyperchaotic systems is done with the initial conditions $x(0) = (0.2, 0.1, 5, 0.1)^T$, $x(0) = (0.5, 0.5, 2, 0.5)^T$, respectively. The values of the constants used for the SMC are $k_1 = k_2 = k_3 = 2$, $\rho_1 = 5$, $\rho_2 = 2$, $\rho_3 = 2$.

For masking, a sinusoidal signal in the form $s = \sin(2\pi 10t)$ and a speech signal available in MATLAB named with "*handle.mat*" are used.

Results of the secure communication with the sinusoidal signal are shown from Figs. 5, 6, 7 and 8. Synchronisations of the system considered as the master system

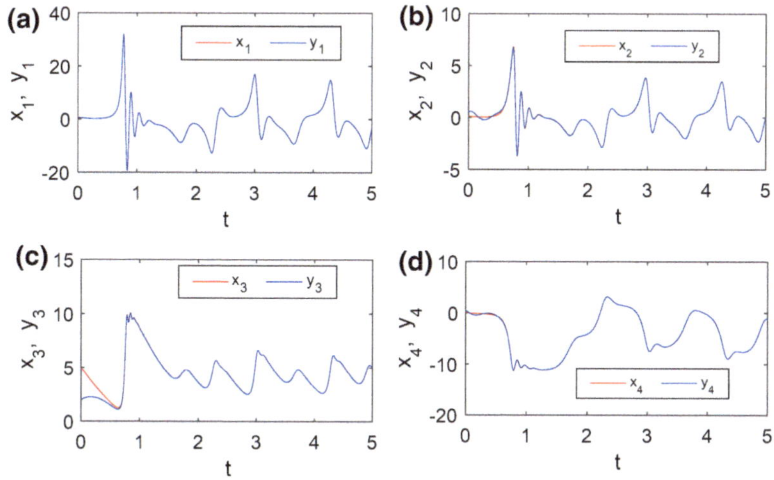

Fig. 5 Synchronised states of (4) and (5) systems with sinusoidal signal

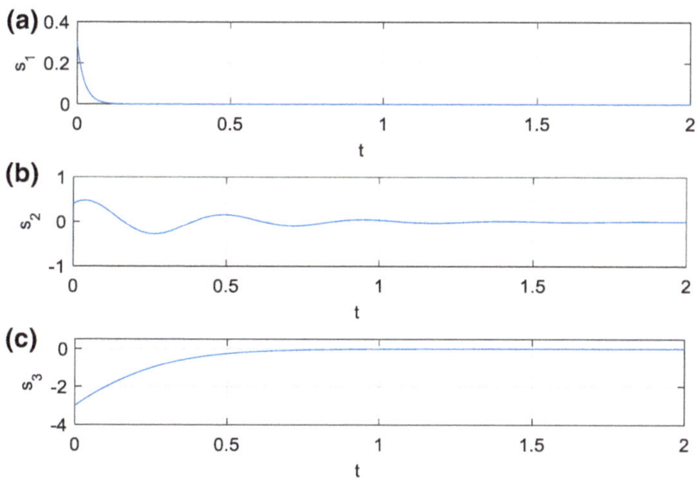

Fig. 6 Behaviour of the sliding surface designed for the synchronisation with sinusoidal signal

and system considered as the slave systems having synchronised states are shown in Fig. 5. It is apparent from Fig. 5 that the states of the master and slave systems are synchronised properly. The time behaviours of the designed sliding surfaces and the designed control inputs are shown in the Figs. 6 and 7 respectively. The nature of the carrier, masked and the transmitted signals along with recovered signal are presented in Fig. 8. Figure 8 discussed that the transmitted message signal is recovered properly.

Now a sound signal is used for the secure communication. The synchronisation errors between the master and slave systems having synchronised states are shown

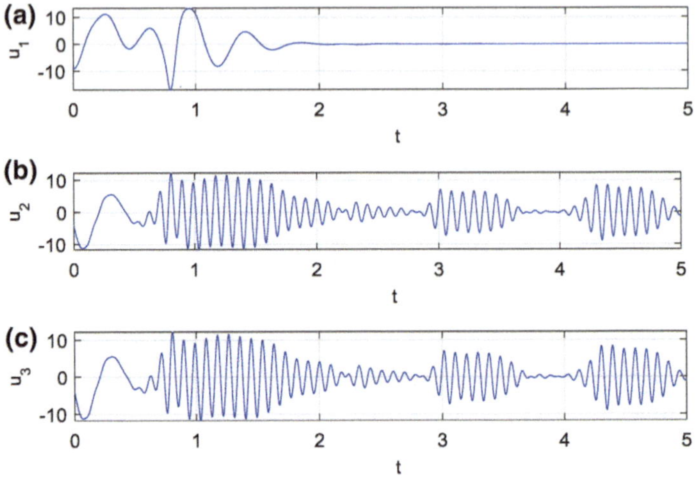

Fig. 7 Behaviour of the designed inputs used for the synchronised system as master (4) and slave (5) systems with sinusoidal signal

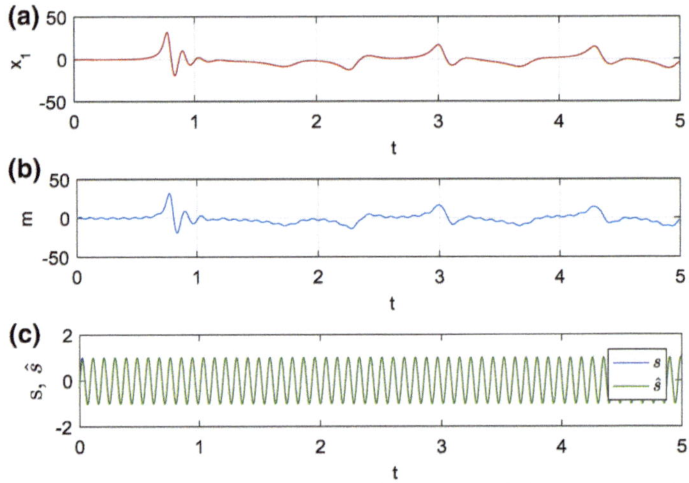

Fig. 8 Responses of the carrier, masked and the transmitted sinusoidal message signal along with recovered sinusoidal message signal

in Fig. 9. It is apparent from Fig. 9 that master and slave systems are synchronised properly in smaller synchronising time. Behaviour of the transmitted message signal, masked signal and recovered message signal in case sound wave is shown in Fig. 10. It is seen from Fig. 10 that the transmitted message signal is recovered properly. Thus, the concept of application in secure communication using the proposed system is validated.

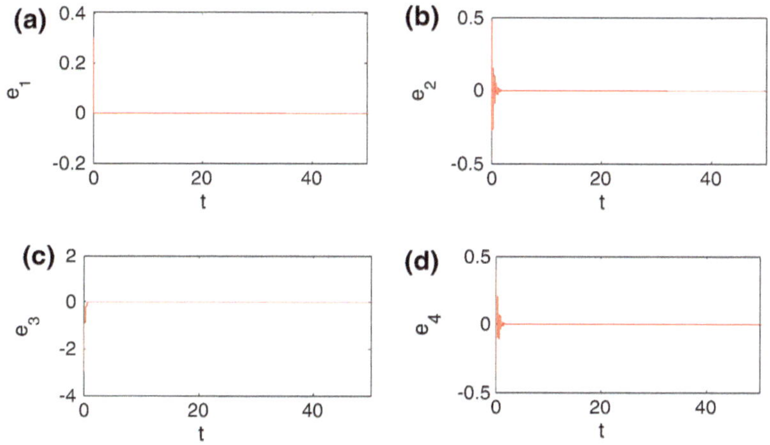

Fig. 9 Synchronisation error between the synchronised states of master (4) and slave (5) systems with considered sound wave message signal

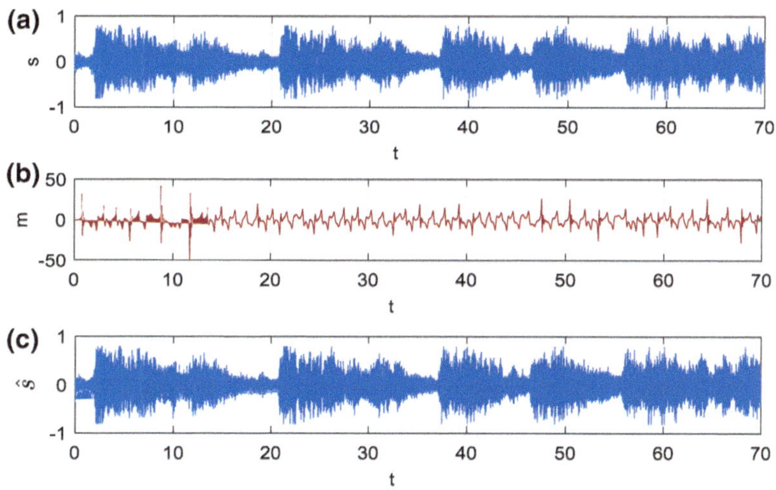

Fig. 10 Responses of the transmitted sound wave message signal, masked signal and recovered message signal

6 Conclusions

In the present work, a new hyperchaotic system is reported. Presence of an equilibrium point having stable nature in the new system makes it to be considered under hidden attractors dynamical system. Dynamical properties in the new system is shown using some numerical methods like phase portrait, Poncaré map and Lyapunov spectrum. The applications of the new system are shown in secure communication by masking of a sinusoidal signal and a sound wave. A PI-SMC is designed for the application. Results using the MATLAB simulation validate the numerous dynamical characteristics and application of the new system.

References

1. Leonov GA, Kuznetsov NV, Kuznestova OA, Seledzhi SM, Vagaitsev VI (2011) Trans Syst. Control 6(2):1–14
2. Klokov A, Zakrzhevsky MV (2011) Int J Bifurc Chaos 21:2825–2535
3. Leonov GA, Kuznetsov NV, Vagaitsev VI (2011) Phys Lett A 375(23):2230–2233
4. Leonov GA, Kuznetsov NV (2013) Int J Bifurc Chaos 23(1):1330002–130071
5. Leonov GA, Kuznetsov NV, Vagaitsev VI (2012) Phys D 241(18):1482–1486
6. Leonov GA, Kuznetsov NV, Kiseleva MA, Solovyeva EP, Zaretskiy AM (2014) Nonlinear Dyn 77(1–2):277–288
7. Leonov GA, Kuznetsov NV, Mokaev TN (2015) Commun Nonlinear Sci Numer Simul 28(1–3):166–174
8. Leonov GA, Kuznetsov NV, Mokaev TN (2015) Eur Phys J Spec Top 224(8):1421–1458
9. Lorenz EN (1963) J Atmos Sci 20(2):130–141
10. Chen G, Ueta T (1999) Int J Bifurc Chaos 9:14651999
11. Lü J, Chen G, Cheng D, Celikovsky S (2002) Int J Bifurc Chaos 12(12):2917–2926
12. Singh JP, Roy BK (2015) Int J Control Theory Appl 8(3):1005–1013
13. Singh PP, Singh JP, Roy BK (2014) Chaos Solitons Fractals 69:31–39
14. Singh JP, Roy BK (2015) Int J Control Theory Appl 8(3):1015–1023
15. Singh JP, Roy BK (2016) Chaos Solitons Fractals 92:73–85
16. Singh JP, Roy BK (2016) Optik 127(24):11982–12002
17. Singh JP, Lochan K, Kuznetsov NV, Roy BK (2017) Nonlinear Dyn 90(2):1277–1299
18. Wei Z, Zhang W, Wang Z, Yao M (2015) Int J Bifurc Chaos 25(2):1550028
19. Wei Z, Zhang W (2014) Int J Bifurc Chaos 24(10):1450127
20. Akgul A, Calgan H, Koyuncu I, Pehlivan I, Istanbullu A (2016) Nonlinear Dyn 84(2):481–495
21. Singh JP, Roy BK (2017) Int J Dyn Control 6:529–538
22. Jafari S, Sprott JC (2013) Chaos Solitons Fractals 57:79–84
23. Li Q, Zeng H, Li J (2015) Nonlinear Dyn 79(4):2295–2308
24. Pham V-T, Jafari S, Volos C, Giakoumis A, Vaidyanathan S, Kapitaniak T, Trans IEEE (2016) Circuits Syst II Express. Briefs 63(9):878–882
25. Singh JP, Roy BK (2017) Optik 145:209–217
26. Singh JP, Roy BK (2017) Nonlinear Dyn 89(3):1845–1862
27. Singh JP, Roy BK, Jafari S (2017) Chaos Solitons Fractals 106:243–257
28. Dudkowski D, Jafari S, Kapitaniak T, Kuznetsov NV, Leonov GA, Prasad A (2016) Phys Rep 637:1–50
29. Kingni ST, Pham V-T, Jafari S, Kol GR, Woafo P (2016) Circ Syst Signal Process 35(6):1807–1813
30. Kingni ST, Jafari S, Pham V-T, Woafo P (2017) Math Comput Simul 132:172–182

31. Wei Z, Moroz I, Wang Z, Sprott JC, Kapitaniak T (2016) Int J Bifurc Chaos 26(7):1650125
32. Li C, Sprott JC, Thio W, Zhu H (2014) Trans IEEE Circ Syst II Express Briefs 61(12):977–981
33. Zhusubaliyev ZT, Mosekilde E, Churilov AN, Medvedev A (2015) Eur Phys J Spec Top 306:6–15
34. Wang L, Yang XS (2015) IFAC-PapersOnLine 48(18):205–210
35. Pham V-T, Volos C, Kapitaniak T (2017) In: Systems with hidden attractors, pp 21–35
36. Wang X, Pham V, Jafari S, Volos C, Munoz-pacheco JM, Tlelo-cuautle E (2017) IEEE Access 5:8851–8858
37. Wang X, Chen G (2011) In: Proceedings—4th international workshop on chaos-fractals theories and applications, IWCFTA 2011, pp 82–85
38. Pham V, Jafari S, Kapitaniak T, Volos C, Kingni ST (2017) Int J Bifurc Chaos 27(4):1750053–1750061
39. Jafari MA, Mliki E, Akgul A, Pham V-T, Kingni ST, Wang X, Jafari S (2017) Nonlinear Dyn 88(3):2303–2317
40. Kingni ST, Jafari S, Simo H, Woafo P (2014) Eur Phys J Plus 129(76):1–16
41. Wei Z, Wang Z (2013) Kybernetika 49(2):359–374
42. Molaie M, Jafari S, Sprott JC, Golpayegani SMRH (2013) Int J Bifurc Chaos 23(11):1350188
43. Wang X, Chen G (2012) Commun Nonlinear Sci Numer Simul 17(3):1264–1272
44. Wei Z, Yang Q (2011) Nonlinear Anal Real World Appl 12(1):106–118
45. Wei Z, Pehlivan I (2012) Optoelectron Adv Mater Rapid Commun 6(7–8):742–745
46. Wei Z (2012) Comput Math Appl 63(3):728–738
47. Wei Z, Yang Q (2011) Nonlinear Dyn 68(4):543–554
48. Huan S, Li Q, Yang X-S (2013) Int J Bifurc Chaos 23(1):1350002–1350006
49. Wei Z, Yu P, Zhang W, Yao M (2015) Nonlinear Dyn 82(1–2):131–141
50. Wei Z, Moroz I, Sprott JC, Akgul A, Zhang W (2017) Chaos 27(3)
51. Wolf A, Swift JB, Swinney HL, Vastano JA (1985) Physics D 16(3):285–317
52. Wu X, Wang K, Wang X, Kan H (2017) Nonlinear Dyn 90(2):855–875
53. Ge X, Lu B, Liu F, Luo X (2017) Nonlinear Dyn 90(2):1141–1150
54. Abd MH, Tahir FR, Al-Suhail GA, Pham V-T (2017) Nonlinear Dyn 90(4):2583–2598
55. Andrievskii BR, Fradkov AL (2004) Autom Remote Control 65(4):505–533
56. Bartolini G, Fridman L, Pisano A, Usai E (2008) Modern sliding mode control theory. Springer, Berlin, Heidelberg

Manhattan Distance Based Voronoi Partitioning for Efficient Multi-robot Coverage

Vishnu G. Nair and K. R. Guruprasad

Abstract In this paper we address the problem of area coverage using multiple cooperating robots. One of the main concerns of using multiple robots is of avoiding repetitive coverage apart from complete coverage of the given area. Partitioning the area to be covered into cells and allotting one each cell to each of the robots for coverage is a simple and elegant solution for this problem. However, the spacial partitioning may lead to additional problems leading to either incomplete coverage or coverage overlap near the partition boundary. We propose a manhattan distance based Voronoi partitioning scheme of $2D \times 2D$ gridded region, where D is the size of the robot footprint. We show that the proposed partitioning scheme completely eliminates coverage gaps and coverage overlap using illustrative results.

1 Introduction

The problem of coverage of a region of interest using autonomous robots has many applications such as in autonomous vacuum cleaning, lawn moving, land mine detection [1], etc. In these applications, the mobile robot is required to move through the region of interest such that every point in this space has come under a coverage tool (or sensor, in applications such as land mine detection) at least once. Apart from this completeness of coverage, the coverage path is also expected to avoid coverage overlap or repeated coverage. Several algorithms have been proposed in the literature to solve this problem both when the area to be covered is known a priori (off-line algorithms) and not known a priori (online algorithms). A survey of various coverage algorithms is provided in [2] and [3].

V. G. Nair
Department of Aeronautical and Automobile Engineering, Manipal Institute of Technology, Manipal Academy of Higher Education, Manipal 576104, Karnataka, India

K. R. Guruprasad (✉)
Department of Mechanical Engineering, National Institute of Technology, Surathkal, Karnataka, India
e-mail: krgprao@gmail.com

C. Shreesha and R. D. Gudi (eds.), *Control Instrumentation Systems*, Lecture Notes in Electrical Engineering 581,
https://doi.org/10.1007/978-981-13-9419-5_7

Using multiple robots to cover a specified region is expected to reduce the coverage time, apart from possible robustness to failure of a (or a few) robot(s). However, a challenging problem when using a Multi-Robotic System (MRS) is to achieve distributed coordination between different robots so that they do not collide with each other movement or perform repeated coverage of regions within the environment. Several multi-robot coverage (MRC) strategies have been proposed in the literature [4–9]. In most of these algorithms, fundamentally two strategies are followed to ensure cooperation and avoid coverage overlap. In the first approach, the regions already covered by all the robots are stored in a central location, and the robots continuously communicate with the central information provider ensuring an indirect cooperation. This approach, apart from increased communication overhead and increased spatial (memory) complexity, may not be suitable in situation where a central information provider cannot be used. In the second approach, each robot should not only keep track of the region it has already covered, in order to avoid self-redundant coverage, but also needs to communicate the covered region to other robots in the team. This results in higher communication overhead along with high memory requirement in each robot.

A simple and elegant technique to reduce the communication requirement is to use divide and conquer approach. Here, the region to be covered is divided into cells and each robot is allotted a cell or a group of preferably contiguous cells for coverage. With this no communication between robots is required while performing coverage. Further, each robot solves a single robot coverage algorithm. In [10, 11] authors decompose the region into cells and dynamically allocate a cell (considered a task) to a robot, until all the cells are covered. In [12] authors divide the region to be covered into n polygons, where n is the number of robots, and each robot is allowed to cover exactly one cell. While in most partition and cover approaches, the region is decomposed into cells by a central computer, and not taking into account the current robot positions, authors in [8] propose a Voronoi Partition-based Coverage (VPC) strategy, where the robots partition the region to be covered into Voronoi cells, considering their current location as nodes.

Though the partition and cover approach solves problems associated with cooperation between robots, and eliminates the on-the-go communication requirement, partitioning itself results in reduced coverage performance in terms of incomplete coverage and coverage overlap at single robot level. In this paper we propose a Manhattan distance based Voronoi partitioning scheme in a 2D $\times$ 2D gridded region to eliminate incomplete coverage and coverage overlap due to presence of cell boundary.

2 Problem Setting

The robot is assumed to have a square footprint of sides D. A robot covers a region while it moves along a path. Typically, whenever possible, robot moves in a straight line. Thus, effectively the swept region will be almost same irrespective of the exact

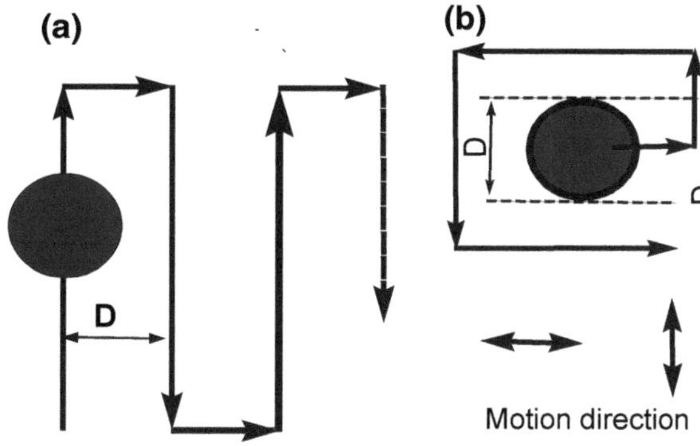

Fig. 1 Typical robot path during coverage **a** back and forth or Booustrophedon path and **b** spiral path. In both cases the robot motion is restricted either in up-down or in right-left directions

shape of the coverage tool (circle, line, etc.) as long as it has width D. The size of the footprint is associated with the coverage tool/sensor rather than the physical size of the robot itself, though, typically, the coverage tool size is comparable to that of the robot itself.

Most coverage algorithms use simple back and forth motion, as illustrated in Fig. 1a or spiraling motion as illustrated in Fig. 1b, as these directions are most effective while covering a free space up/down and right/left. In certain algorithms, a wall following algorithm may be used [13–15] to circumnavigate an obstacle in order to achieve a truly complete coverage. As we focus on incomplete or repetitive coverage induced by partition boundary, and not the physical boundary of the region, or the presence of obstacles, we do not consider robot motion apart from in up/down and right/left directions. Note that it is not possible to eliminate the incomplete or repetitive coverage problems arising due to physical boundary of the region and the presence of obstacles in arbitrary situations.

Restriction on robot motion direction also leads us to decomposition of the region to be covered into square cells of size $D \times D$. Let us call a $D \times D$ cell as a sub cell. A sub cell is covered only if it is completely free of obstacles or completely inside the region to be covered. Thus, we assume that a partially occupied (by obstacle) cell or a cell partially inside the region to be covered remains uncovered. Note that this is not a restrictive assumption, as if such cells need to be covered, wall following path needs to be created and it will lead to coverage overlap. The coverage is said to be complete if all completely free mi-nor cells are visited once by the robot (that is, resolution complete), and non-repetitive, if no free sub cell is visited more than once. Further, we assume that the region to be covered is free from obstacles. Though this assumption might seem unrealistic, we do not consider obstacles, as the focus of this paper is only on the effect of partitioning on coverage performance. The proposed partition and cover approach can handle obstacles like most of the coverage algorithms reported in the literature such as [13, 16]. When obstacles are present, the single-robot coverage algorithm generates a coverage path avoiding the obstacle.

The problem addressed in this paper is to devise a partitioning scheme which will eliminate incomplete or repetitive coverage induced by partition boundary, and use it to devise a multi-robot coverage strategy using a partition and cover approach.

3 The Proposed "Partitioning and Cover" Scheme

In this section we present the proposed "partition and cover" strategy for multi-robot coverage problem addressed here. First we shall discuss how the partition affects the performance of a coverage algorithm.

3.1 Partition Boundary Induced Issues

Consider a scenario illustrated in Fig. 2. The region shown here can be covered by a single robot completely and without any overlap/retraced path. Now let us partition the region into two cells and each cell is allotted a robot to accomplish coverage. The main purpose of using multiple (two in this case) robots is to reduce the time required to cover entire area. As each robot has to cover a smaller area now, coverage time is reduced. However, the partition boundary (shown as thick solid line) passes through a few sub cells (shaded) and splits them into two parts, one each on partitioned cells. Coverage path of the robots is shown in Fig. 2 using solid lines with arrows indicating the direction of motion. The grid lines decomposing the region into D $\times$ D sub cells are shown by dashed lines. The gridding is shown by dashed lines. It can be observed that neither of the robots covers the shaded sub cells, leading to incomplete coverage. Note that, if we force the robots to cover these split (shaded) sub cells, the robot path should be along the partition boundary, and this will lead to coverage overlap. Apart from incomplete coverage, the robot retraces its path at a few instances. These are indicated by circled paths in Fig. 2. Note that the same region is covered completely by a single robot without any overlap as shown in Fig. 3.

3.2 Partitioning Scheme to Avoid Path Retrace

Now we present the proposed partitioning scheme to achieve the objective of eliminating the partition induced incomplete or repetitive coverage problems. It can be observed that the robot is forced to retrace the path on entering certain sub cells as there is no free return path for the robot. A path of width 2D is required if a return path has to be accommodated. Now if instead of D $\times$ D gridded space, let us consider a 2D $\times$ 2D gridded space as illustrated in Fig. 4. The 2D $\times$ 2D grids are shown by thick dashed lines, while thin dashed lines show D $\times$ D gridding. Each 2D $\times$ 2D cell

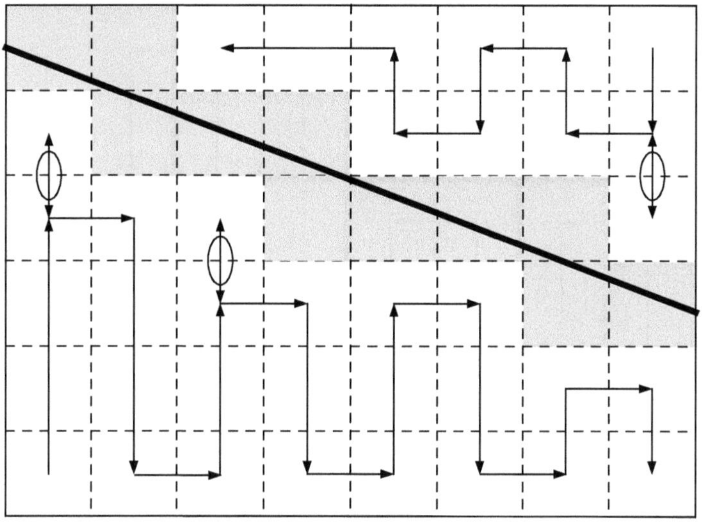

Fig. 2 A partition boundary (shown as thick solid line) in continuous space leads to coverage gap and/or coverage overlap. The grids are shown in dashed lines, while solid lines with arrow show robot path. Uncovered regions (cells) are shown in grey. Robot retraces path in cells circled leading to coverage overlap

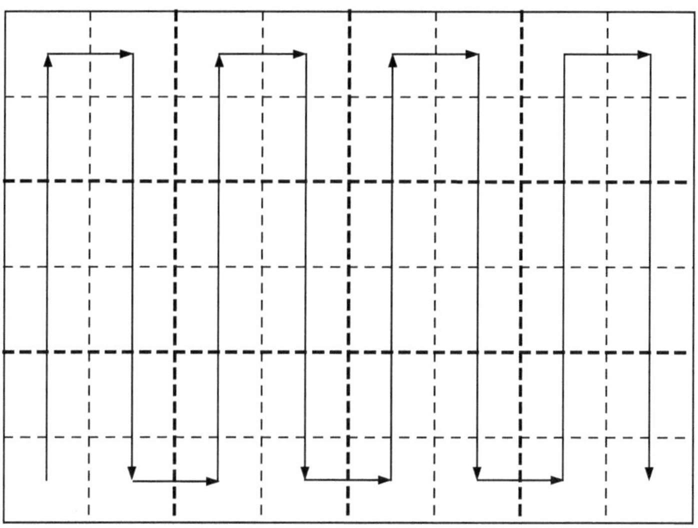

Fig. 3 A single robot completely covers the region without any path retrace or coverage overlap

is made up of four sub cells. Let us call a 2D × 2D cell as a major cell, a terminology used in [13].

Now let us partition this 2D × 2D gridded space as illustrated in Fig. 4. Here, the partition boundary is shown with thick line. Now we can observe that both robots

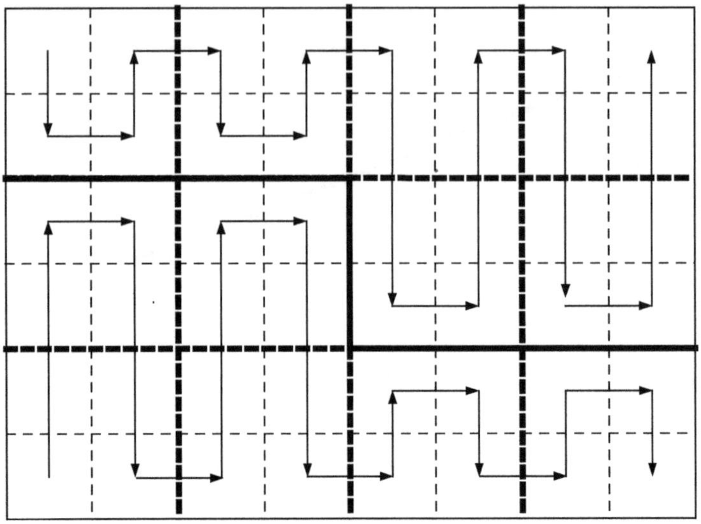

Fig. 4 A partition in 2D × 2D gridded space (boundary shown as thick solid line) eliminates both the coverage gap and coverage overlap. The grids (2D × 2D) are shown in thick dashed lines, while solid lines with arrow show robot path

together cover all sub cells without any retrace/overlap. Thus, we can eliminate the partition induced coverage inefficiency completely by partitioning the 2D × 2D gridded region.

In all the scenarios, we have used back and forth (Boustrophedon) path for the purpose of illustration. It is easy to verify that any coverage path (such as spiraling motion) with robot motion restricted to up/down and left/right will lead to similar result. Further, we have used a simple scenario for illustrating the problem of incomplete and overlapping coverage with decomposition scheme in the continuous space and D × D gridded space, and how these partition-induced problems can be eliminated using partitioning in a 2D × 2D gridded space.

3.3 Coverage Path and Manhattan Distance

As we have mentioned in the problem setting, in most coverage path planning algorithms, the robot moves in either up/down (*Y* direction) or left/right (*X* direction), at least in free space. This directional restriction may be relaxed only around the obstacle boundaries or the boundary of the region to be covered (if a truly complete coverage is desirable, at the cost of coverage overlap). In fact, restricting motion along only *X* or *Y* directions in free space leads to improved coverage completeness and reduces (or eliminates) coverage overlap.

Now as the robot can move only in either X or Y directions, it makes sense to measure distance between any two pints in the space too along X direction and Y direction. This leads us to a metric known as Manhattan distance. Consider two points $P_1 = (x_1, y_1)$ and $P_2 = (x_2, y_2)$. Now the Euclidian distance between these two points is given by,

$$d(P_1, P_2) = \sqrt{(x_1 - x_2)^2 + (y_1 - y_2)^2} \quad (1)$$

The corresponding distance measured using Manhattan metric is

$$d_m(P_1, P_2) = |(x_1 - x_2)| + |(y_1 - y_2)| \quad (2)$$

For a robot having to move along either X or Y (not both simultaneously),the distance it needs to travel to reach P_2 starting at P_1 is $d_m(P_1, P_2)$, the Manhattan distance, rather than the Euclidean distance. Thus it makes sense to use Manhattan distance in case of a coverage path planning problem.

3.4 Manhattan Distance Based Voronoi Partitioning of 2D x 2D Gridded Space

Voronoi partitioning [17] has been widely used as an effective spatial partitioning tool in many applications including coverage optimization in multi-agent(robotic) systems [18] and [8]. A standard Voronoi partitioning scheme decomposes a space using the concept of nearness to nodes. Let $I_N = \{1, 2, \ldots, N\}$; $Q \subset \mathbb{R}^2$; and $\mathcal{P} = \{p_1, p_2, \ldots, p_N\}$, $p_i \in Q$, be a set of points in Q called a *node set*. The *Voronoi partition*, generated by $\mathcal{P}$ is the collection $\{V_i(\mathcal{P})\}_{i \in I_N}$ with,

$$V_i(\mathcal{P}) = \left\{q \in Q \mid \parallel q - p_i \parallel \leq \parallel q - p_j \parallel, \forall j \in I_N\right\} \quad (3)$$

The Voronoi cell V_i is the collection of those points which are closest to p_i. In the context of the multi-robot coverage problem addressed in this paper, N is the number of robots, p_i is the position of the ith robot, Q is the region to be covered.

As we have discussed in previous section, in the context of coverage problem addressed here, the Manhattan distance metric is more suitable than the standard Euclidean metric. From the perspective of robot travel distance, "closeness" of any two points is measured in terms of the Manhattan distance. Thus, it makes sense to replace Euclidean distance in Eq. (3) by the Manhattan distance. Now we have Manhattan distance-based Voronoi partition given by,

$$V_i(\mathcal{P}) = \left\{q \in Q \mid d_m(q, p_i) \leq d_m(q, p_j), \forall j \in I_N\right\} \quad (4)$$

3.5 Partitioning a Gridded Space

Note that the Manhattan-distance based Voronoi partitioning scheme given in Eq. (4) partitions the continuous space Q. Now instead if we chose Q_D to be the $D \times D$ gridded (discretized) region Q, to be collection of all gridded cells in it, that is,

$$Q_D = \{c, \text{a girdded cell} | c \subset Q\} \tag{5}$$

Now we may partition Q_d into Manhattan distance based Voronoi partition as,

$$V_{Dmi}(\mathcal{P}) = \left\{c_i \in Q_d | d_m(c_i, p_i) \leq d_m(c_i, p_j), \forall j \in I_N\right\} \tag{6}$$

Here, c_i is the centroid of the gridded cell c. Thus V_{di} is collection of all gridded cells, whose centroid is closest to p_i, in Manhattan sense.

3.6 The Manhattan-VPC Strategy

With this background we now present the proposed Manhattan Voronoi partition based coverage (Manhattan-VPC), a strategy using the Manhattan distance based Voronoi partition. As every point within a Voronoi cell is closest to the corresponding robot, it makes sense to allot each robot to cover the Voronoi cell associated with it [8]. In this paper, in place of the standard Voronoi partition, we use (a generalized) Voronoi partition of the $2D \times 2D$ gridded space, Q_{2D}, using Manhattan distance, d_m. We denote such a partition by V_{2Dm}, and the ith generalized Voronoi cell as V_{2Dmi}.

Let $\mathcal{P} = \{p_1, p_2, \ldots, p_N\}$, $p_i \in Q$, the location of N robots. A Voronoi partition of Q_{2D} is created using the Manhattan distance and the ith robot covers the region V_{2Dmi}. Individual robot may use any single robot coverage algorithm reported in the literature for covering the corresponding (generalized) Voronoi cell.

4 Results and Discussions

In this section we provide an illustrative simulation result to demonstrate the proposed "partition" and "cover" strategy. We considered Spanning Tree Coverage (STC) [13], a coverage algorithm based on approximate cellular decomposition, as representative single robot coverage strategies. The simulation is carried out in Matlab at the graph-level. We have considered coverage using three robots.

Figure 5 shows the coverage path using proposed Manhattan-VPC of the $2D \times 2D$ gridded space, with STC algorithm as the underlying single robot coverage strategy. As it can be observed from the result, the coverage is complete and without any overlap in both cases. That is, a complete and non-overlapping coverage is achieved

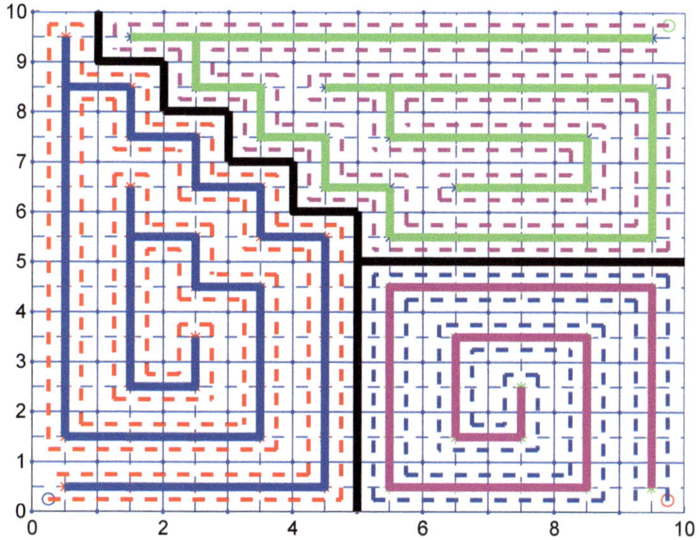

Fig. 5 Coverage by three robots using STC algorithm with Manhattan distance based Voronoi partition using 2D × 2D grids. The colored (grey) solid lines show the spanning tree and the actual robot path through minor cells is shown by dashed lines

irrespective the single-robot coverage algorithm used,as long as the underlying single robot coverage algorithm guarantees to provide complete/non overlapping coverage.

5 Conclusion

We proposed a "partition" and "cover" strategy for cooperative multi-robot coverage, using Voronoi partitioning scheme based on Manhattan distance metric in a gridded region is discussed in the chapter. The region is divided into $2D \times 2D$ grids, where $D \times D$ is the robot (coverage tool) footprint. This gridded region is partitioned using Manhattan distance-based Voronoi partitioning scheme. With the help of illustrative example, we demonstrated that the proposed partitioning scheme eliminates partition boundary induced incompleteness and overlap in coverage, using existing single robot coverage strategies. A detailed formal analysis and experimental investigation is part of the ongoing work.

References

1. Dasgupta P, Muoz-Melndez A, Guruprasad KR (2012) Multi-robot terrain coverage and task allocation for autonomous detection of landmines. Proc SPIE 8359
2. Choset H (2001) Coverage for robotics—a survey of recent results. Ann Math Artif Intell 31(1–4):113–126

3. Galceran E, Carreras Marc (2013) A survey on coverage path planning for robotics. Robot Auton Syst 61:1258–1276
4. Agmon N, Hazon N, Kaminka (2009) The giving tree: constructing trees for efficient offline and online multi-robot coverage. In: Annals of math and artificial intelligence
5. Michel D, McIsaac K (2012) New path planning scheme for complete coverage of mapped areas by single and multiple robots. In: IEEE international conference on mechatronics and automation, pp 1233–1240
6. Hazon N, Kaminka G (2011) On redundancy, efficiency, and robustness in coverage for multiple robots. Robot Auton Syst 56:1102–1114
7. Zheng X, Koenig S, Kempe D, Jain S (2010) Multirobot forest coverage for weighted and unweighted terrain. IEEE Trans Robot 26:1018–1010
8. Guruprasad KR, Wilson Z, Dasgupta P (2012) Complete coverage of an initially unknown environment by multiple robots using voronoi partition. In: Proceedings of 2nd international conference on advances in control and optimization of dynamical systems
9. Wilson Z, Whipple T, Dasgupta P (2011) Multi-robot coverage with dynamic coverage information compression. In: 8th international conference on informatics in control, automation and robotics, pp 236–241
10. Jager M, Nebel B (2002) Dynamic decentralized area partitioning for cooperating cleaning robots. In: Proceedings 2002 IEEE international conference on robotics and automation, pp 3577–3582
11. Min T, Yin H (1998) A decentralized approach for cooperative sweeping by multiple mobile robots. In: Proceedings of international conference on intelligent robots and systems
12. Hert S, Lumelsky V (1998) Polygon area decomposition for multiple robot workspace division. Int J Comput Geom Appl 8:437–466
13. Gabriely Y, Rimon E (2001) Spanning tree based coverage of continuous areas by a mobile robot. Ann Math Artif Intell 31:77–98
14. Ranjitha TD, Guruprasad KR (2015) CCSTC: a truly complete competitive spanning tree coverage algorithm for mobile robots. In: Proceedings of advances in robotics, 2nd international conference of robotics society of India. BITS Goa, India, July 2015
15. Ranjitha TD, Guruprasad KR (2016) Truly complete competitive robot coverage path planning using approximate cellular decomposition. In: Proceedings of 4th IFAC conference on advances in control and optimization of dynamical systems ACODS, IFAC-PapersOnLine, vol 49, issue 1, Tiruchirappalli, India, pp 195–200
16. Choset H (2000) Coverage of known spaces: the boustrophedon cellular decomposition. Auton Syst 9:247–253
17. Okabe A, Boots B, Sugihara K, Chiu S (2000) Spatial tessellations. Concepts and applications of Voronoi diagrams. Wiley
18. Guruprasad KR, Dasgupta P (2012) Distributed spatial partitioning of an initially unknown region for a multi-robot coverage application. In: Autonomous robots and multrobot systems (ARMS) workshop (co-located with AAMAS 2012)

Deposition of ZnO Thin Film at Different Substrate Temperature Using RF Sputtering for Growth of ZnO Nanorods Using Hydrothermal Method for UV Detection

Basavaraj S. Sannakashappanavar, C. R. Byrareddy, Sanjit Varma, Nandini A. Pattanshetti and Aniruddh Bahadur Yadav

Abstract The growth properties of ZnO Nanorods was studied on different seed layers by the deposition of Zinc oxide (ZnO) thin film on SiO_2/Si substrate by RF sputtering at two different conditions, i.e., one at room temperature and another at 400 °C substrate temperature. Surface morphology of the seed layer was studied by X-Ray Diffraction and Atomic Force Microscopy (AFM). Low cost hydrothermal method was employed for the growth of ZnO Nanorods on both the seed layers. The structural properties of ZnO nanorods were characterized by Field Emission Scanning Electron Microscope (FESEM). The FESEM images showed the proper alignment and orientation of ZnO nanorods grown on both the seed layers. The I-V measurements were carried out at room temperature under dark light and Ultraviolet (UV) light source. In order to examine the UV detection, MSM (Metal–Semiconductor–Metal) photodetector was fabricated and responsivity was measured for the nanorods grown on both seed layers. The better responsivity and contrast ratio of ZnO nanorods based UV detector was observed in case of 150 nm seed layer deposited at 400 °C.

Keywords Seed layers · ZnO nanorods · Hydrothermal method · UV detection

B. S. Sannakashappanavar (✉)
Department of Electronics & Telecommunication, ADCET, Ashta 416301, Maharashtra, India
e-mail: raj.ec008@gmail.com

C. R. Byrareddy
Department of Electronics & Communication, BIT, Bangalore 560004, Karnataka, India

N. A. Pattanshetti
Department of Chemistry, Karnataka University Dharwad, Dharwad 580003, Karnataka, India

S. Varma · A. B. Yadav
Department of Electronics & Communication, SVEC, Tirupati 517102, Andra Pradesh, India

C. Shreesha and R. D. Gudi (eds.), *Control Instrumentation Systems*,
Lecture Notes in Electrical Engineering 581,
https://doi.org/10.1007/978-981-13-9419-5_8

1 Introduction

ZnO exhibits excellent optical properties and therefore has proved to be a potential material for short wavelength applications such as, photodetectors (PDs), lasers, and light-emitting diodes. ZnO posses a wide band gap (3.37 eV) and large excitation binding energy of 60 meV [1, 2] at room temperature. It is also considered as a promising metal oxide semiconductor material for short wavelength applications like ultraviolet (UV). The Ultraviolet (UV) photodetectors are used in a wide range of applications, like, in space research, ozone layer monitoring, missile warning systems, high temperature flame detection, environmental monitoring, etc. [3, 4].

In recent days one-dimensional (1D) nanostructures have been widely studied because of their promising applications in nanoelectronic devices such as field-effect transistors [5], photodiodes [6], and chemical sensors [7]. Different methods are used for the fabrication of ZnO nanorods, such as metal organic chemical vapor deposition (MOCVD), pulsed laser deposition (PLD), hydrothermal method [8] etc. Among these the cost effective hydrothermal method was employed which has fewer limitations as compared to other methods.

The surface morphology, alignment, orientation, and optical properties of the ZnO nanorods depends on precursors, surface treatment and heat treatment which plays a significant role in affecting all the above properties.

In this proposed work, two different ZnO thin films were used as a seed layers fabricated at two temperatures, i.e., one at room temperature and the other at an elevated temperature of 400 °C. This acts as different precursors for nanorod development. Further, the ZnO nanorods were fabricated on these two seed layers using hydrothermal method. To investigate the effect of precursor on quality of grown ZnO nanorods AFM, FESEM and XRD analysis were performed. Electrical properties were studied by I-V measurement. And finally, UV detection properties were examined upon the obtained samples.

2 Experimental

2.1 *Materials Used*

All the chemicals were purchased from Sigma Aldrich (France) and were used without any further purification. Zinc nitrate hexahydrate ($Zn(NO_3)_2.6H_2O$) and Hexamethylenetetramine (HMTA) ($C_6H_{12}N_4$) were used for the growth of ZnO Nanorods. A p-type Silicon wafer with <100> orientation and 4–7 Ω-cm resistivity was used as substrate material. For the entire experimental process Deionised distilled water was used with 18.2 MΩcm resistivity.

2.2 Methods

Preparation of ZnO seed layer

A p-type silicon wafer with <100> orientation was cleaned using RCA1 and RCA2 cleaning technique and dried for further use. Silicon dioxide (SiO_2) was deposited by thermal oxidation (Dry oxidation) method at 1100 °C for about 2 h on silicon (Si) in order to obtain SiO_2/Si substrate. The thickness of SiO_2 was measured to be ~200 nm using ellipsometer. In the present work, ZnO thin film acts as seed layer for growth of ZnO Nanorods. ZnO thin film was deposited with two different substrate temperature conditions using RF Sputtering which measures to be ~150 nm on SiO_2/Si substrate. The two samples of RF sputtered ZnO thin films were labeled as sample A and sample B. Sample A was ZnO seed layer deposited at room temperature under argon atmosphere with 5×10^{-5} of base vacuum pressure and Sample B was ZnO seed layer deposited at 400 °C with ~150 nm thicknesses. The Argon gas pressure was maintained around 2.0×10^{-2} mbar. The target to substrate distance was 14 cm, 100 W RF power and 13.56 MHz of RF generator frequency was maintained respectively. Similarly, the sputtering condition for sample B was maintained same as above but the substrate temperature was set to 400 °C. The deposition rate sample A was measured to be 5 nm/min and for sample B was around 3 nm/min. Further, the samples A and B were used for the growth of ZnO nanorods.

Growth of ZnO nanorods

A low-cost hydrothermal method was employed for the growth of ZnO Nanorods using an equimolar (20 mM) solutions of Zinc nitrate hexahydrate ($Zn(NO_3)_2.6H_2O$) and Hexamethylenetetramine (HMTA) ($C_6H_{12}N_4$) in 150 ml DI water [9]. Both the samples were immersed upside down in the beaker containing the above solution. The beaker was placed in hot air oven at 95 °C. After 4 h, the samples were removed and rinsed twice with DI water and dried in air. The schematic illustration of the as grown ZnO nanorods on ZnO/SiO_2/Si substrate is shown in Fig. 1a.

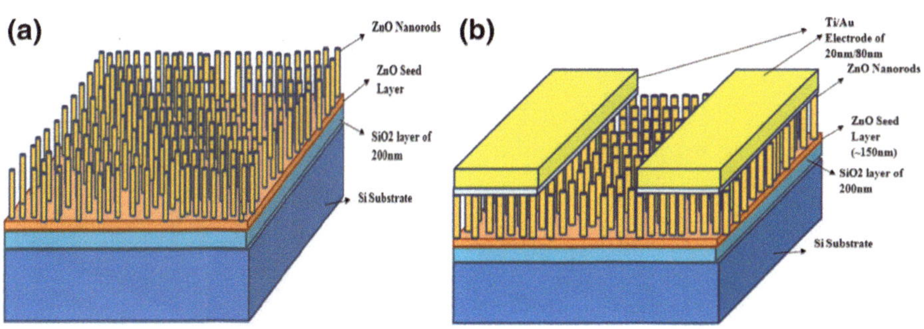

Fig. 1 **a** Schematic of ZnO nanorod growth and **b** device structure

3 Characterization Methods

The thickness of the RF sputtered ZnO thin film and (SiO_2) layer was measured using ellipsometer (make SENTECH Instruments, Model SE 800). The surface topography and structural properties of both the ZnO seed layers was observed using Atomic force microscopy (AFM) (PICO SPM, model PICO scan 210) and X–ray Diffractometer (make RIGAKU, model Smartlab 3KW) respectively. Surface morphologies of ZnO nanorods were examined using Field Emission Scanning Electron Microscopy (make Zeiss Model Ultra 55). The IV characterization study of the vertically aligned ZnO nanorods based photodetector was performed to assess the electrical properties using semiconductor parameter analyzer (Keithley 4200-SCS).

4 Results and Discussion

ZnO seed layers with 150 nm thickness deposited at room temperature was designated as sample A and 150 nm deposited at 400 °C was labeled as sample B. The depositions were carried out by RF sputtering on SiO_2/Si substrate as explained in the preceding sections. Figure 2 represents the 3D AFM images of sample A and B ZnO seed layers. The AFM analysis revealed that the average roughness for sample A was greater (1.81 nm) than compared to sample B (1.19 nm). Further, the RMS (root mean square) roughness values were 2.44 and 2.20 nm for A and B seed layer samples respectively. And Average Height was observed to be 6.9–5.8 nm.

Figure 3 represents the XRD pattern of ZnO seed layer of both the samples. High resolution X-ray diffraction (make Rigaku, model Smartlab 3KW) with CuKα radiation was used with the wavelength of 1.5420 Å for the measurement of crystallite structure. The source and detector was kept constant while measuring the theta-2theta angle. The XRD peaks were compared with the standard JCPDS card which proves that the peaks are in good agreement with a hexagonal wurtzite crystal structure [10].

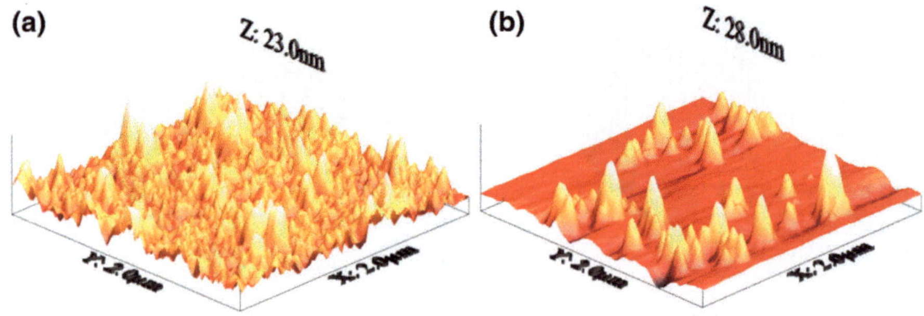

Fig. 2 **a** 3D AFM image of ZnO seed layer Sample A, **b** 3D AFM image of ZnO seed layer Sample B

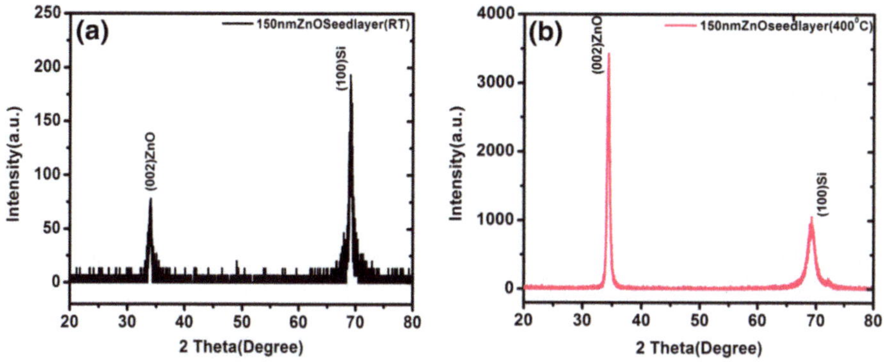

Fig. 3 **a** XRD pattern of ZnO Seed layer Sample A, **b** XRD pattern of ZnO seed layer Sample B

XRD images shows the diffraction peak of ZnO 002 plane at 34.02 for sample A and 34.31 for sample B and Si peak at 69.08 and 69.19 for both the samples respectively. It was observed that intensity of peak increases as substrate temperature increased.

Figure 4 represents the FESEM images of the nanorods grown on both A and B ZnO seed layers samples. From the images a uniform growth of ZnO nanorods was observed on both the samples possessing a hexagonal surface. The diameter of the nanorods grown on sample A ranges between 40–42 nm and for sample B was 63–75 nm respectively. On the other hand, the average length of the nanorods was between 0.94–1.1 μm and 1.2–1.3 μm on the seed layer A and B respectively. According to the AFM analysis the value of roughness and average height of grains was increased in sample A compared to sample B and in XRD pattern the c axis oriented peak intensity was increased in sample B compared to sample A. This proves that as grain size decreases density of the nanorods increases. The effect of seed layer [11] and substrate [12] can be observed here and Also the variation in growth properties of ZnO Nanorods can be observed for different seed layers on SiO_2/Si substrate.

Figure 1b Represents the device structure of our proposed Nanorod based MSM photodetector. The device was fabricated using shadow mask technique [13]; the two electrodes of Ti/Au were deposited on vertically grown ZnO nanorods on both the samples using E-beam evaporator of 20 nm/80 nm thick. The dimensions of shadow mask were measured to be 1050 μm wide and 2000 μm long. The spacing between the two electrodes was 150 μm. The IV measurements were taken for both the samples at room temperature under dark light and UV light source of different wavelengths (365 and 380 nm) as shown in Fig. 5a, b. The responsivity of the Nanorod based UV photodetector was measured using [14–16].

$$R = \frac{I_{ph}}{P_o} \tag{1}$$

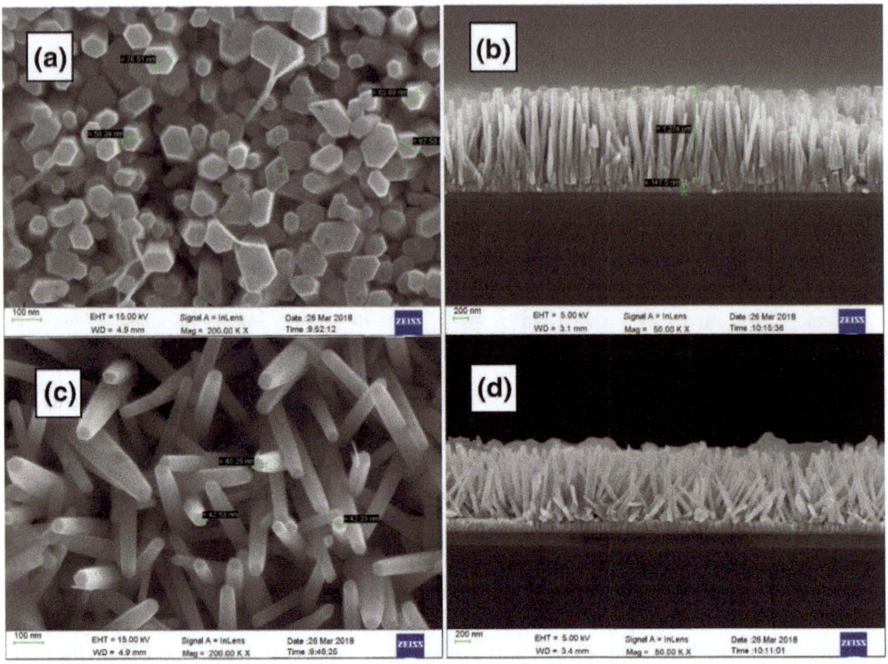

Fig. 4 FESEM image of ZnO nanorods of **a** and **b** 150 nm ZnO seed layer (400 °C) surface and cross-section **c** and **d** 150 nm ZnO seed layer (room temperature) surface and cross-section

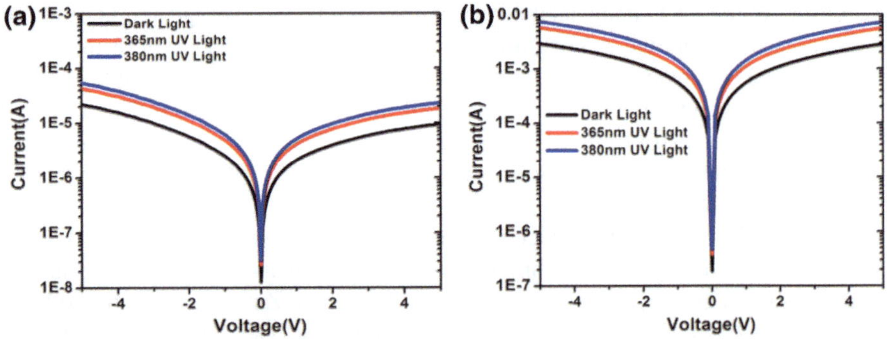

Fig. 5 **a** IV characteristics of Sample A measured in dark light and UV light and **b** IV characteristics of Sample B measured in dark light and UV light

Table 1 Detailed parameters of ZnO Nanorod based MSM UV photodetector device from observed IV Characteristics

ZnO seed layer	UV light (nm)	Photocurrent (A) (at V = +5 V)	Responsivity (A/W)
150nm ZnO (RT)	365	1.8754×10^{-5}	0.00468
150 nm ZnO (RT)	380	2.3442×10^{-5}	0.00586
150 nm ZnO (400 °C)	365	0.00556	1.39
150 nm ZnO (400 °C)	380	0.00723	1.80

where I_{ph} represents the photocurrent (A), and P_o is the power of UV source. The measured values of responsivities of Photodetector is shown in Table 1.

At room temperature, under dark light, when an external potential is applied the absorption of thermal energy is driven by the electrodes, which causes the electrons trapped in ZnO Nanorods can be excited to the conduction band. The photocurrent measured in the device containing of ZnO Nanorods is the photo generated current generated by electron-hole pairs. The holes are captured by the trap levels present at the surface of ZnO Nanorods, because of this the enhancement of photocurrent is observed at the electrodes. In this work, the concentration of growth solution is same in both the samples, but the seed layer deposition was carried out at different temperatures which leds to the different morphologies of ZnO Nanorods (i.e. size and height). It is observed that responsivity is increasing in Sample B compared to sample A. Therefore these devices can be used in wide applications of various electronic and optoelectronic systems.

5 Conclusion

In the present work, the effect of seed layer deposition on the growth of Nanorods was studied. The morphology demonstrated the variation in their UV detection properties using a lab-scale UV detector fabricated by the authors. The comparative study on the growth of ZnO nanorods was carried out by depositing ZnO seed layer at different substrate temperature. The AFM data gave the roughness value and average height of grains which was observed to be higher in sample A compared to sample B. This affected the growth rate and diameter of the nanorods as proved by the FESEM images. In addition, the grown ZnO Nanorods achieved the crystallite hexagonal wurtzite structure in case of both the seed layers as obtained from the XRD data. Thus the obtained results showed that the growth of ZnO nanorods is greatly influenced by the nature of seed layer and the comparative study revealed that the 150 nm seed layer deposited at 400 °C presented a more improved results for UV detection. Finally, ZnO nanorods grown on both seed layers were used to fabricate a MSM photodetector. We have examined the ZnO Nanorod based MSM photodetector against UV light and calculated the responsivity for different seed layers. The above study proves that further optimization of densely packed ZnO Nanorods can be used for fabricating novel electronic devices which can be used for different practical applications.

Acknowledgements The authors would like to thank Indian Institute of Technology, Bombay for accepting the project proposal under the Indian Nanoectronics User Program (INUP). The author also likes to thank the faculties under INUP program for their proper guidance during the experimental work and overall support till the project completion.

References

1. Radzimska AK, Jesionowski T (2014) Zinc oxide—from synthesis to application: a review. Materials 7:2833–2881
2. Chee CY, Nadarajah K, Siddiqui MK, Wong Y (2014) Optical and structural characterization of solution processed zinc oxide nanorods via hydrothermal method. Ceram Int 40:9997–10004
3. Sang L, Liao M, Sumiya M (2013) Sensors 13:10482–10518
4. Tian C, Jiang D, Zhao Y, Liu Q, Houa J, Zhao J, Liang Q, Gao S, Qin J (2014) Mater Sci Eng B 184:67–71
5. Frenzel H, Lorenz M, Lajn A, von Wenckstern H, Biehne G, Hochmuth H, Grundmann M (2009) ZnO-based metal-semiconductor field-effect transistors on glass substrates. Appl Phys Lett 95:153503
6. Yakuphanoglu F, Caglar Y, Caglar M, Lican S (2010) ZnO/p-Si heterojunction photodiode by sol–gel deposition of nanostructure n-ZnO film on p-Si substrate. Mater Sci Semicond Process 13:137–140
7. Shawn Lim S-H, Raorane D, Srinath S, Majumdar A (2006) Nanochemo-mechanical sensor array platform for high-throughput chemical analysis. Sens Actuators B 119:466–474
8. Ikizler B, Peker SM (2014) Effect of the seed layer thickness on the stability of ZnO nanorod arrays. Thin Solid Films 558:149–159
9. Sannakashappanavar BS, Pattanashetti NA, Byrareddy CR, Yadav AB (2018) Study of annealing effect on the growth of ZnO nanorods on ZnO seed layers. AIP Conf Proc 1943:020077
10. Ko YH, Nagaraju G, Yu JS (2015) Fabrication and optimization of vertically aligned ZnO nanorod array-based UV photodetectors via selective hydrothermal synthesis. Nanoscale Res Lett
11. Lee SH, Kim SH, Yu JS (2016) Metal-semiconductor-metal near-Ultraviolet (~380 nm) photodetectors by selective area growth of ZnO nanorods and SiO2 passivation. Nanoscale Res Lett
12. Lajvardi M, Ghazi ME, Izadifard M, Eshghi H, Hadi I (2018) Effect of seed layer thickness on optoelectronic properties of ZnO-NRs/p-Si photodiodes. Optik
13. Sannakashappanavar BS, Byrareddy CR, Kumar PS, Yadav AB (2018) Seed layer effect on different properties and UV detection capability of hydrothermally grown ZnO nanorods over SiO2/p-Si substrate. Superlattices Microstruct 117:503–514
14. Ali GM, Chakrabarti P (2010) ZnO-based interdigitated MSM and MISIM ultraviolet photodetectors. J Phys D Appl Phys 43:415103
15. Singh S, Park S-H (2015) Fabrication and characterization of Al:ZnO based MSM ultraviolet photodetectors. Superlattices Microstruct
16. Sannakashappanavar BS, Byrareddy CR, Pattanashetti NA, Singh K, Yadav AB (2019) Growth of ZnO nanorods on different seed layer thickness using the hydrothermal method for UV detection. J Nanoelectron Optoe 14:964–971

3D Printable Modules for Manually Reconfigurable Manipulator with Desired D-H Parameters

Doddabasappa Marebal and K. R. Guruprasad

Abstract Modular robots are designed to increase the utilization of robots by modularizing their architecture. We discuss manually reconfigurable manipulators, where a manipulator of desired kinematic configuration is built by assembling the available modules. In the case of a serial-link manipulator with revolute joints, the joint angle is a variable. Out of the remaining three D-H parameters, namely, link-length, link-offset and link-twist, the twist angle influences the workspace the most. This work proposes a conceptual design and fabrication of individual modules which can be assembled to obtain a modular manipulator with desired kinematic configuration in terms of twist angles between any two consecutive joints. We also discuss possible provisions for length adjustment of a link. Designed modules are fabricated using 3D printer. As we focus on manually reconfigurable manipulators, simplicity of individual modules, in terms design, fabrication, and assembly, has been given higher priority, in contrast to similar designs available in the literature.

Keywords Manipulator · Modular robots · Manually reconfigurable robots

1 Introduction

Robotic manipulators find applications in many fields such as industrial automation, defense, surgery and other medical fields, service, research, etc. One of the primary task a manipulator needs to perform in any of the applications is to move the end effector or the tool in a desired path. A manipulator consists of several links connected by joints which allow constrained motion between the links. The joints

D. Marebal
Department of Automation and Robotics, KLE Technological University, Hubballi, India
e-mail: doddabasappa@bvb.edu

K. R. Guruprasad (✉)
Department of Mechanical Engineering, National Institute of Technology, Surathkal, Karnataka, India
e-mail: krgprao@gmail.com

C. Shreesha and R. D. Gudi (eds.), *Control Instrumentation Systems*, Lecture Notes in Electrical Engineering 581,
https://doi.org/10.1007/978-981-13-9419-5_9

used in manipulators generally allow one degree of freedom. Common joints used are revolute, prismatic, spherical, screw and cylindrical. Serial-link manipulators are most commonly used in most applications. Further, owing to their ease of actuation, revolute joints are preferred over other kinds of joints. In this paper, we focus on serial link manipulators with revolute joints.

The Denavit Hartenberg (D-H) parameters are used to describe the kinematic configuration of a manipulator in terms of arrangement of links and relative position and orientation of the links and joint axes. These parameters are link length, joint angle, link twist, and link offset. As the kinematic configuration and the hence the workspace of a manipulator depends on the D-H parameters, kinematic design of the manipulator involves selection of the D-H parameters, along with the number of degrees-of-freedom. Though a robot is multi-functional machine required to perform a wide variety of tasks, in many practical situations, it is designed to perform a set of few tasks depending on the application. Desired workspace of a manipulator is decided based on the task it needs to perform. Hence the task to be performed dictates the selection of the D-H parameters.

1.1 Modular Robots

A robot with fixed D-H parameters has limited capabilities in terms of the kind of tasks it can perform. The manipulator may have to be replaced if the task to be performed changes significantly and the current manipulator is no more suitable for the new task. As part of process of kinematic design of a manipulator, prototype of each configuration designed (in terms of D-H parameters) may have to be built to test and validate the design. In the context of robotic research, as kinematics, dynamics, control, of a manipulator depend on the D-H parameters, multiple manipulators with different D-H parameters may be required to experimentally validate any new concept, such as a new control law. In most these situations, instead of building or procuring multiple robots, it makes sense to build a manipulator with desired D-H parameters, by assembling general purpose low cost modules. Modular robotics addresses this problem. Here a robot is built by assembling the available modules.

In this work, we address a problem of designing 3D printable individual modules for a manually configurable modular robot. We provide a simple design of modules, which can be assembled manually to obtain a serial-link manipulator having revolute joints with desired D-H parameters, namely, link twist angles and link lengths.

1.2 Literature Review

Modular robots may be manually configured or self configurable. Self reconfigurable modular robots [1, 3] are self governing kinematic machines with variable morphology, structure and usefulness. Substantial amount of work has been carried out in the field of self-reconfigurable modular robots. Most such modular robots form into chains to enhance motion capability, rather than into a manipulator. Manually configurable modular robotic manipulators are very useful in several applications. In the following, we preview representative works from the literature on manually reconfigurable robotic manipulators.

A design concept called Toshiba Modular Manipulator System (TOMMS) has been proposed by Matsumaru [2] to achieve a modular manipulator system. TOMMS consists of joint modules, link modules, and a control unit with a joystick. Joint module can be used only in two configurations, namely, vertical and horizontal, limiting the possible kinematic configurations in terms of twist angles. The link lengths are fixed. Schonlau [4] presented Modular Manipulator System (MMS), a general purpose user configurable manipulator system. It contains two types of modules namely joint modules and link modules. Two types of joint modules are provided one for linear motion and the other rotational motion (Prismatic and revolute joints). Link modules are either a straight link or an elbow link. The number of configurations (in terms of twist angle) is limited in this design. Further, the link lengths are fixed.

Singh et al. [6] proposed a modular concept with provision for changing twist angle and link lengths in steps. Link length adjustment is achieved by a simple telescopic arrangement. End of the links are joints which can be rotated to select the desired twist angle. In a more recent work [5], authors present an adaptive unit using worm and half spur gears, for twist angle adjustment. Though this design enables automatic change of twist angle, it makes the unit more complex in terms of fabrication and maintenance. Both designs are conceptual in nature and authors do not provide any details of physical realization. Also, provision for placement of joint actuators and controller are not discussed.

Manual reconfigurable robots are very useful in many applications where a user can build a manipulator based on the requirement of the task. Such a system will also be a very useful tool as an aid in teaching and research. Though there has been substantial reported work on self-reconfigurable modular robots, not much work on manually configurable modular robots is available in the literature. Further, the designs of manually reconfigurable modular robots reported in the literature are either conceptual and complex, when providing sufficient flexibility in terms of possible twist angles and link lengths [5, 6], or provide limited flexibility when simple in nature and complete in deigns, including actuation and control, as provided in [2] and [4]. In this work we attempt to conceptualize, design and fabricate 3D printable modules that are simple yet provide sufficient flexibility in terms of twist angle and link length adjustment.

2 Conceptual Design of Modules

In this section we discuss the conceptual design of the proposed modular robots. First we provide a discussion on importance of each D-H parameter on the manipulator workspace.

2.1 Effect of D-H Parameters on the Workspace

Out of four D-H parameters, the joint angle, link offset, link length, and link twist, either the joint angle or the link offset are variables. In case of a prismatic joint, joint angle θ_i is a fixed parameter, while link offset d_i is the joint variable. In the case of a revolute joint, the joint angle θ_i is the joint variable and link offset (d_i), link twist (α_i), and link length (a_i or l_i) are fixed parameters. Since we restrict our attention to a serial manipulator with revolute joints, the D-H parameters of interest are link offset (d_i), link twist (α_i), and link length (a_i or l_i). Though all three parameters affect the kinematic configuration, we may observe that the link twist affects the workspace the most, in the sense that the shape of the workspace is primarily determined by the link twists. Link lengths affect only the size of the workspace rather than its shape. Further, the link offset typically does not affect the workspace in shape or size. This is illustrated in Fig. 1 for a simple two link manipulator with revolute joint.

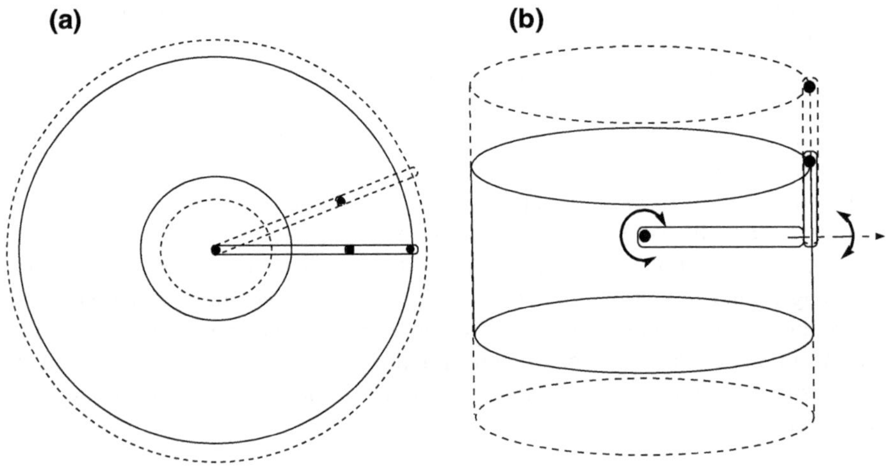

Fig. 1 Workspace of a 2R manipulator with different D-H parameters. **a** A planar configuration with $\alpha = 0°$ has an annular region between two circles. Annular regions between circles shown solid line and dashed lines show workspace with different lengths of the second link. **b** Workspace of non-planar configuration ($\alpha = \pi/2$) is curved surface of a cylinder. Again only size of the workspace changes with link length, shown using solid and dashed lines with different second link length

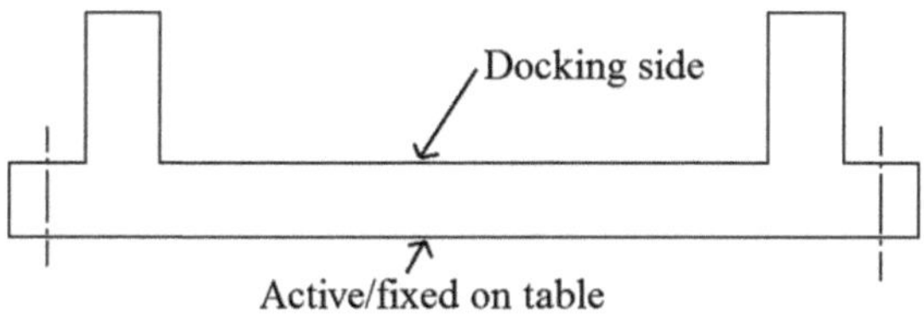

Fig. 2 Schematic view of a base module

As the link twist influences the workspace the most, we consider link twist adjustment as an important feature in the proposed design. Though the link lengths affect the workspace only in terms of its size, it is still an important parameter to be considered in manipulator design. Hence, we primarily focus on provision for link twist adjustment. We also discuss possible solutions for incorporating link length adjustment and provide a simple design for the same.

2.2 Conceptualized Modules

Each module has a docking or attaching side, and an active side. While two modules are attached at the docking sides, the active side may either be fixed or allow a relative motion forming joints. Further, a joint (active side) may be passive or active (or actuated). Different types of modules proposed are, base module, link module, and tool module. The tool module is also called a last link module. Link modules are again of two types, male and female modules. Now we shall discuss each module concept.

Base Module This module has a docking side which may be attached to a male (or female) link module. The active side is fixed on a table. Figure 2 shows schematic view of a base module.

Female link module This module can attach to any other module on its dock side. The active side is an actuated revolute joint, and can only be attached to the active side of a male link module. The concept of a female link module is illustrated in Fig. 3a.

Male link module This module too can be attached to any modules at its docking side. The active side is a passive joint which can be attached only to the active side of a female link module. Figure 3b illustrates a male link joint.

Last link (or tool) module The last link connects to a male (or female) link module on its docking side and a tool such as gripper can be attached to its active side. A conceptual last link module is shown Fig. 4.

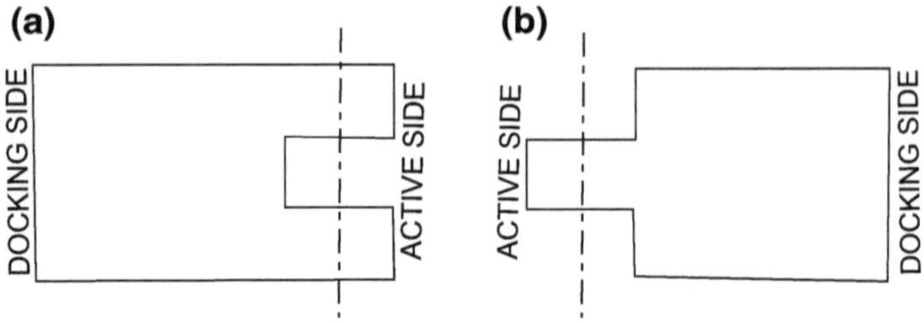

Fig. 3 Schematic of **a** a female link module and **b** a male link module

Fig. 4 Schematic of a last link module

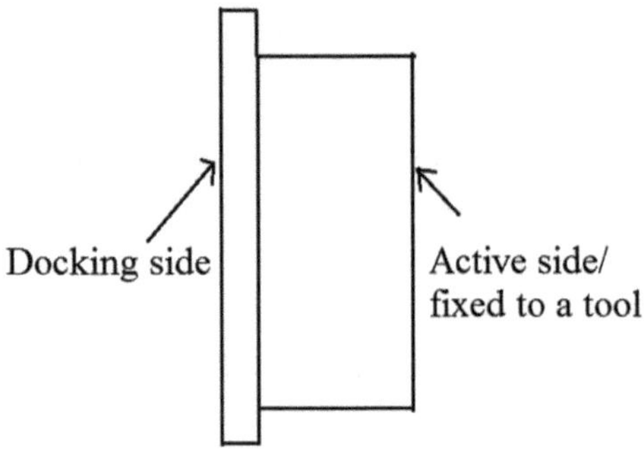

Fig. 5 A base link is formed by attaching a base module to a female module

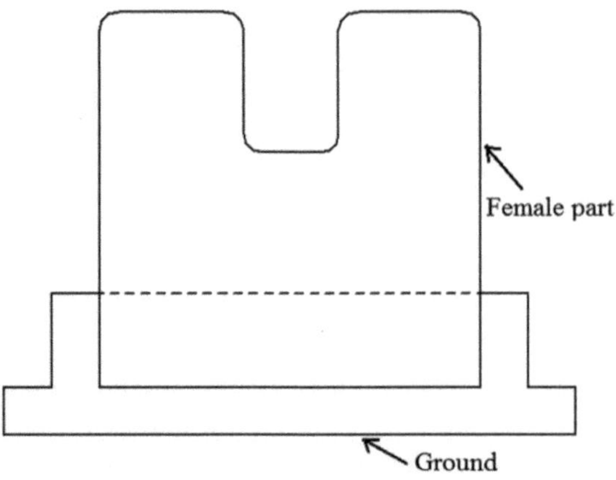

2.3 Assembly of Modules

Now we shall discuss how various components of a manipulator are obtained by attaching a pair of modules.

Base link A base link is obtained when a base module is attached to a female (or male) link as illustrated in Fig. 5. The active side of the base link is fixed to a table.

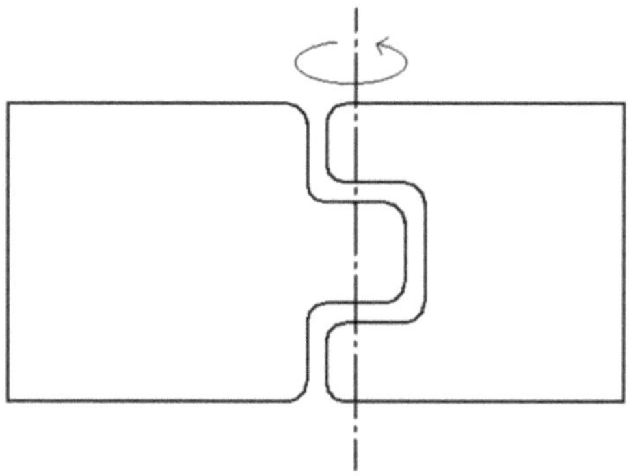

Fig. 6 An actuated joint formed by a male and a female module

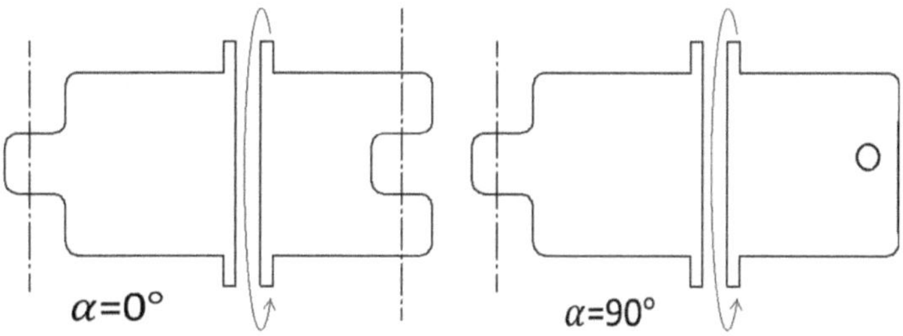

Fig. 7 An intermediate link with desired twist angle is formed by attaching male and female modules at the active sides

An actuated joint An actuated revolute joint is obtained when a male link and a female link modules are attached on their active sides. Figure 6 shows a joint formed by attaching male and female modules.

Intermediate link An intermediate link is obtained by attaching a male link module to a female link module at the docking side rather than the active side, as shown in Fig. 7. Further, the twist angle may be varied by rotating one of the modules as shown in the figure.

Last link A last link is obtained by attaching a last link (tool) module to a male (or female) link module. Figure 8 illustrates a last link obtained by attaching a male link to a tool link. A tool such as gripper can be attached to the active side of the last link module.

2.4 *Provision for Link Length Adjustment*

Though from the perspective of workspace shape, link length may not be as important as the twist angle, in many situations it may be desired to build a manipulator with desired size of workspace, which depends on link lengths. There are several ways in which this can be achieved.

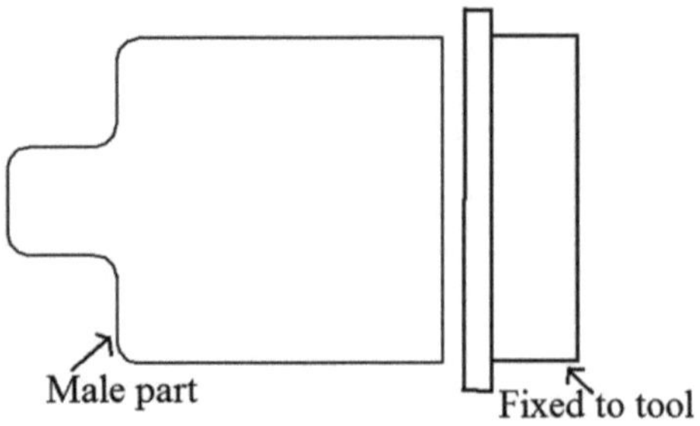

Fig. 8 A last link with provision for attaching a tool

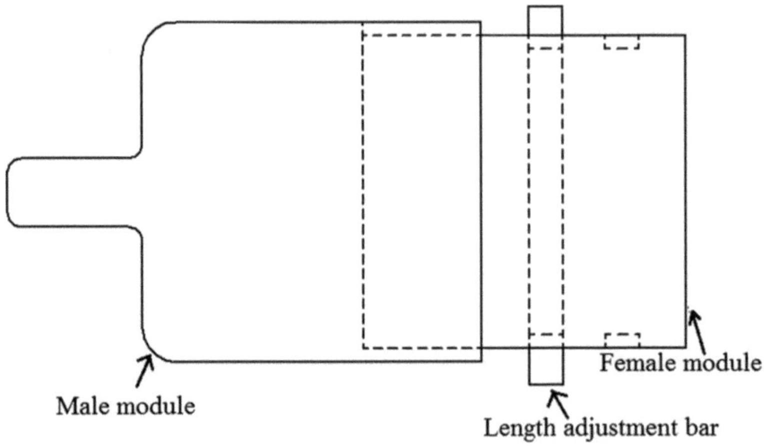

Fig. 9 Provision for small adjustment in link length

1. As proposed in [6], a telescopic mechanism, may be used.
2. A link module may be used, as in TOMMS [2] and MMS [4], along with provision for varying length of the modules, which is not provided TOMMS or MMS. Different link modules with different lengths or telescopic link modules may be used.
3. We propose modified telescopic mechanism here for a small change in link length in steps. The concept is illustrated in Fig. 9.

3 Designed Modules

Based on the modular robot concept discussed in the previous section, we have designed and modeled individual modules using CATIA modeling software. In this section we provide the details of the modules which are ready for fabrication. The

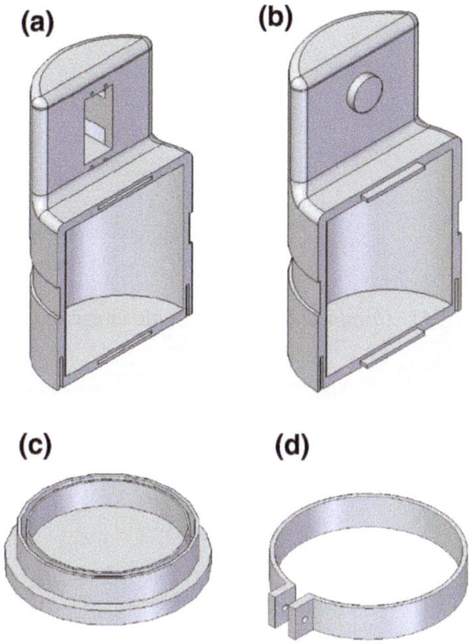

Fig. 10 Components of an integrated base-female link module. **a** Female link module part 1, **b** female link module part 2, **c** base link module, and **d** base link holding clip. A slot is provided in the female module (as shown in (**a**)) for mounting a servo motor to actuate the joint

designs are provided to support the new concept of 3D printable modular robot, we skip dimensional details of the components and provide only pictorial representations.

3.1 Base Link

Instead of a separate base module, we have integrated a base link with the docking female link module as a base link can only dock with a female link module. Figure 10 shows various components of an integrated base-female link module. Figure 10a, b show left and right parts of the female link module used exclusively for the base link. Figure 10c shows the actual base link module and Figure 10d shows a clip for holding the base link. Figure 11 shows an assembled module forming a base link.

3.2 Link Modules

As we have discussed in the previous section, a link is formed by attaching two modules called the male and female link modules. These modules are designed such that on the docking surface they can be assembled to form a link with desired twist angle, and on active sides they form an actuated joint. A provision for fixing a motor

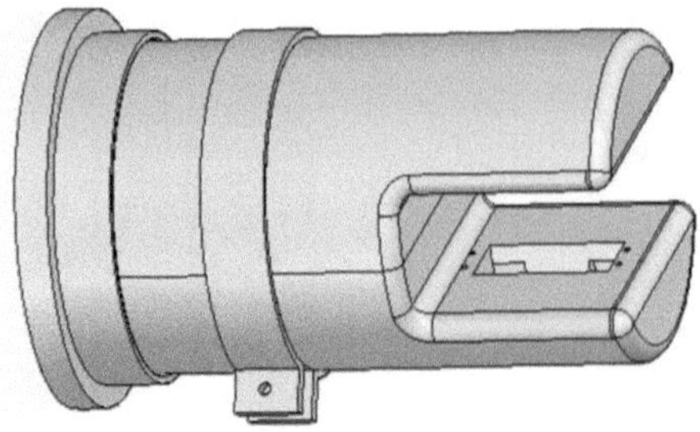

Fig. 11 Complete base link module integrated with a female link module

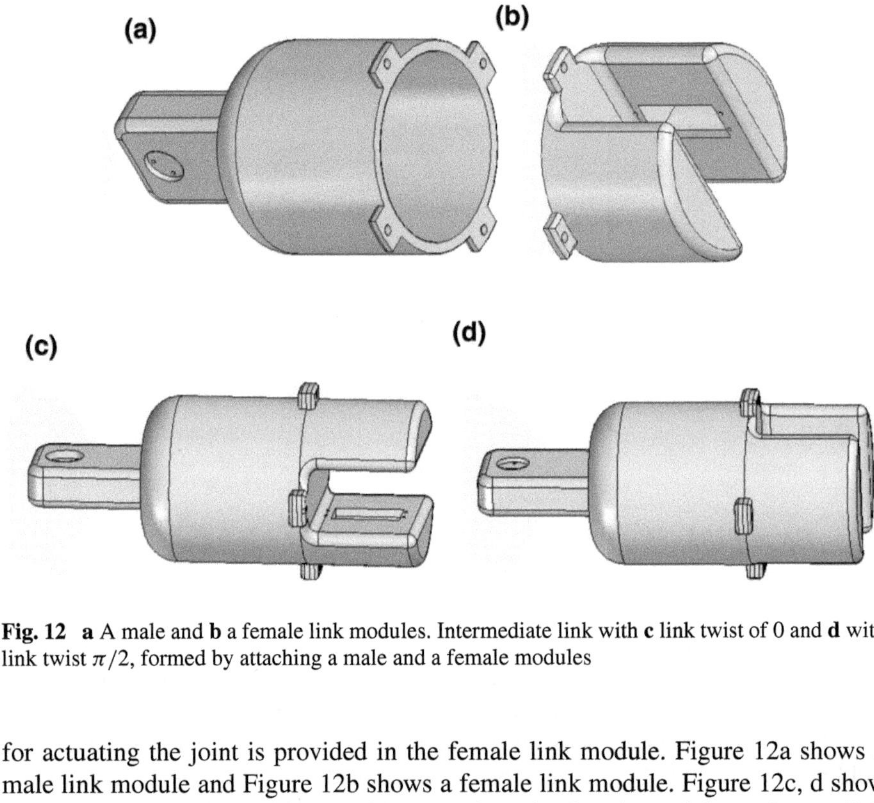

Fig. 12 **a** A male and **b** a female link modules. Intermediate link with **c** link twist of 0 and **d** with link twist $\pi/2$, formed by attaching a male and a female modules

for actuating the joint is provided in the female link module. Figure 12a shows a male link module and Figure 12b shows a female link module. Figure 12c, d show intermediate links formed by attaching a male and a female modules to form a link with zero twist and a link with twist of $\pi/2$, respectively. Though in this design, we have restricted the twist angles in step of 90°, it is very easy to reduce the step size.

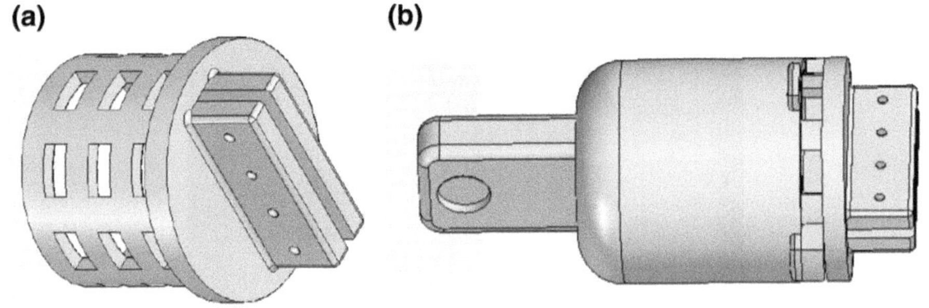

Fig. 13 **a** Last link (tool) module with provision for variable link length, and **b** a last link (tool) module docked to a male link module to form the last link

Fig. 14 Modules assembled to form a simple 2R manipulator

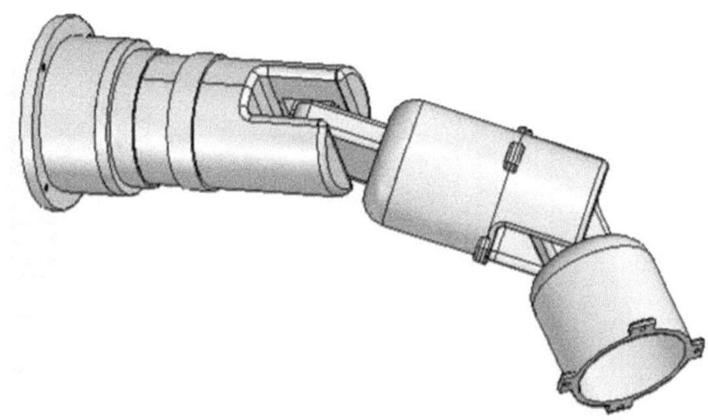

3.3 Last Link Or Tool Module

The last link needs to be attached to a tool on one side (active side) and to a male (or female) link module on the docking side. Figure 13 shows a last link module along with a provision for changing link length in steps. This provision for changing link length can be provided in all female (or male, not both) link modules. As we mentioned earlier, the main focus of this work is to adjust the twist angle, and we have incorporated a provision for length adjustment as shown in Fig. 13a only for the illustration of the concept. Figure 13b shows a tool module attaching to a male module to form a last link. Provision is made so that a tool such as gripper may be attached to the active side of this module. Figure 14 shows a simple 2R manipulator obtained by assembling various modules.

4 Fabrication of the Modules Using 3D Printing

The modules designed are fabricated using a 3D Printer. Material used to print the component is ABS (Acrylonitrile Butadiene Styrene) Filament 1.75 mm.

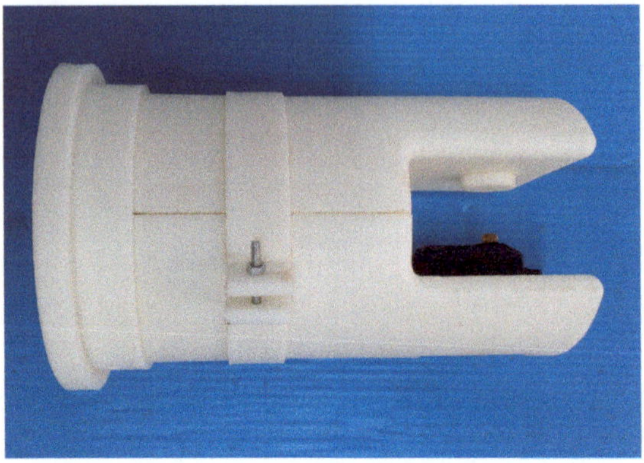

Fig. 15 Photograph of a base link module assembled with a female link module to form a base link

(a) (b)

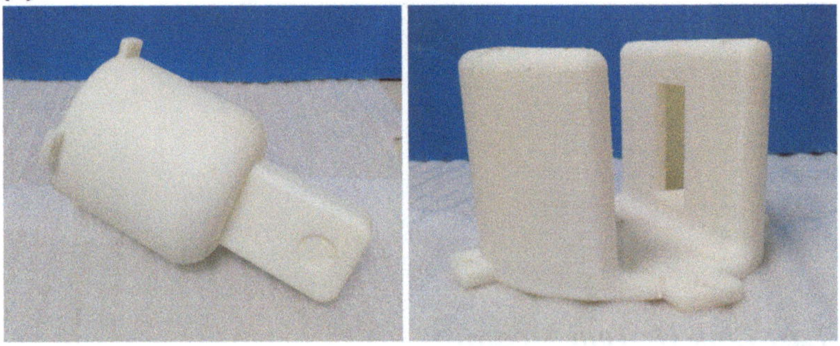

Fig. 16 3D printed **a** male link module and **b** a female link module. Space for a servo motor is provided in the female link module. The figures are not to same scale

Fig. 17 Assembly of 3D printed modules to obtain a last link

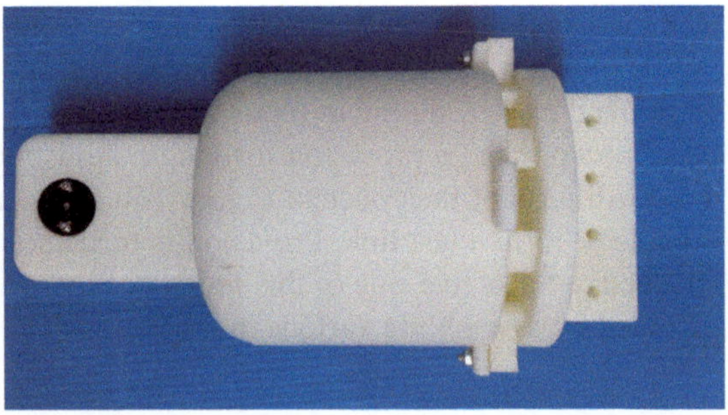

Figure 15 shows a photograph of a base link formed by assembling 3D printed base link module and female link modules. Note that a servo motor has been attached to the female part of the link. Figure 16a, b show 3D printed male and female link modules respectively.

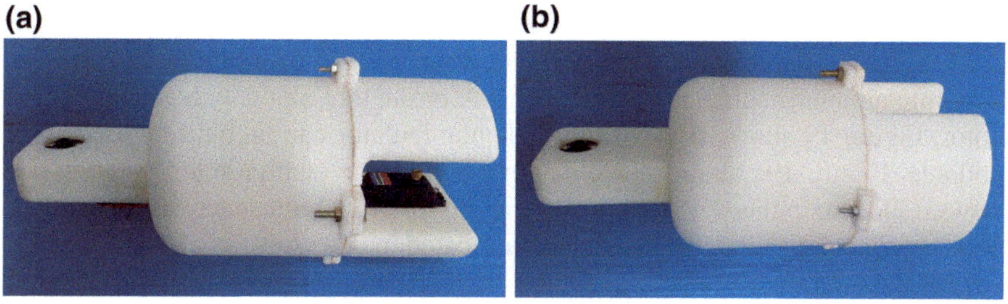

Fig. 18 Assembly of 3D printed modules to obtain an intermediate link with twist angle **a** 0 and **b** $\pi/2$

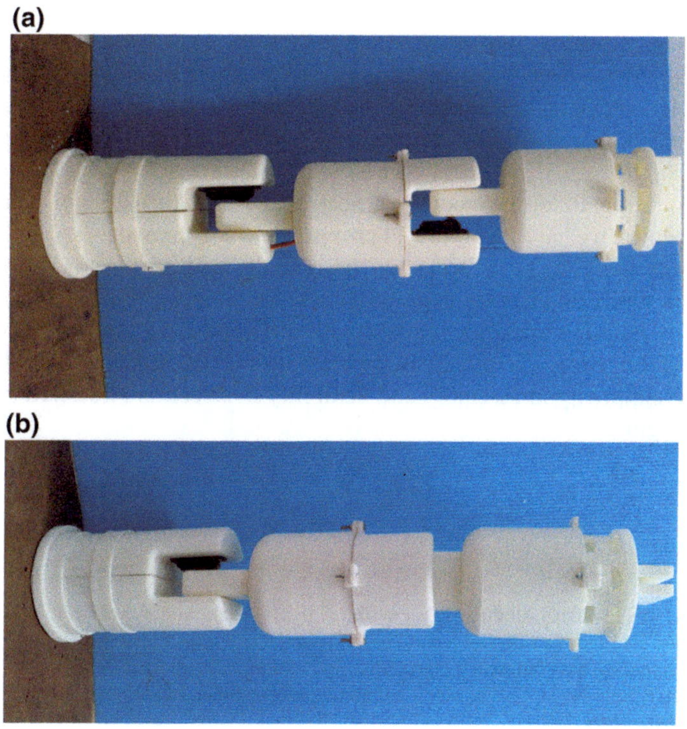

Fig. 19 Assembly of modules to build a 2R manipulator: **a** a planar configuration and **b** a non-planar configuration, along with servo motors to actuate the joints

Figure 17 shows a last link obtained by assembling a male link module with a last link module. A provision for changing link length is provided in the last link module as discussed in the previous section.

Figure 18a, b show intermediate links, with different twist angles, assembled using individual male and female modules, demonstrating the proposed modular robot concept.

Finally, we demonstrate the process of assembling various links obtained by attaching corresponding modules to build a simple 2R manipulator which has a base link, an intermediate link, and a last link or the tool along with actuating servo motor. Figure 19 shows how a simple 2R manipulator can be built by assembling modules. Figure 19a shows a 2R manipulator planar configuration and Fig. 19b shows a 2R manipulator with non-planar configuration, built successfully using the modules conceptualized, designed, and fabricated in this work, by changing the link twist of the intermediate link.

5 Conclusions

In this paper, we have proposed a new simple 3D printable modular robot concept. As the link twist influences the workspace the most, we focused primarily on twist angle adjustment. We have also discussed how link length adjustments can be incorporated into our design.

After conceptualizing the modules, we carried out a detailed design. While designing the modules, we have incorporated provisions to accommodate servo motors and controllers. Finally we have fabricated the modules and demonstrated that a serial manipulator with desired twist angle and link lengths can be built using the available modules, by building a 2R manipulator with different configurations. The detailed design and fabrication is primarily used to demonstrate the proposed modular design, in terms of simplicity in assembly and flexibility in selection of desired twist angle and link lengths.

References

1. Gilpin K, Rus D (2010) Modular robot systems. IEEE Robot Autom Mag 17(3):38–55 (2010)
2. Matsumaru T (1995) Design and control of the modular robot system: TOMMS. In: Proceedings of IEEE international conference on robotics and automation, pp 2125–2131
3. Moubarak P, Ben-Tzvi P (2012) Modular and reconfigurable mobile robotics. Robot Auton Syst 60(12):1648–1663
4. Schonlau W (1997) A modular manipulator systems (MMS) architecture and implementation. In: Proceedings of 8th international conference on advanced robotics, pp 901–906
5. Singh S, Aggarwal A, Singhal Y, Singla E (2015) Development of reconfigurable serial manipulators using parameters based modules. In: Proceedings of 2nd international and 17th national conference on machines and mechanisms iNaCoMM
6. Singh S, Gupta S, Singla E (2012) Design of module for modular customized manipulators. In: Proceedings of the ASME international mechanical engineering congress & exposition

FIR Filter Design Technique to Mitigate Gibb's Phenomenon

Niyan Marchon and Gourish Naik

Abstract A unique methodology employing a linear phase finite impulse response (FIR) low pass filter (LPF) was proposed with an attempt to mitigate passband and stopband ripples due to Gibb's phenomenon. The three regions of the filter response in the frequency domain are approximated using trigonometric functions. The proposed filter model achieved a sharp transition of 2π, fairly flat passband and a stopband attenuation of 40 dB. Our algorithm suppressed the oscillations near the edge of the transition region as well as in the passband region, reducing the Gibb's phenomenon from the conventional passband ripples from 18% to as low as 2%. Thus a three-fold satisfactory performance was achieved in all the three bands namely passband, transition and stopband. Our proposed linear phase FIR LPF was effectively used to filter out power line interference and higher unwanted frequencies from the real time electroencephalogram signals.

Keywords Finite impulse response · Linear phase · Low pass filter · Gibb's phenomenon · Electroencephalogram

1 Introduction

To design a finite impulse response (FIR) filter we can approximate the frequency response H (ω) of filter by calculating its impulse response h (n) [1].

$$H(\omega) = \sum_{n=-\infty}^{\infty} h(n)\, e^{-j\omega n} \quad (1)$$

N. Marchon (✉)
Padre Conceicao College of Engineering, Verna, Goa, India
e-mail: niyanmarchon@gmail.com

G. Naik
Electronics Department, Goa University, Taleigao, Goa, India

C. Shreesha and R. D. Gudi (eds.), *Control Instrumentation Systems*,
Lecture Notes in Electrical Engineering 581,
https://doi.org/10.1007/978-981-13-9419-5_10

$$h(n) = \frac{1}{2\pi}\int_{-\pi}^{\pi} H(\omega)\, e^{j\omega n} d\omega \quad -\infty \leq \mathrm{n} \leq \infty \tag{2}$$

As the duration of the impulse response is infinite and it can be truncated at n = N − 1, to obtain an FIR filter at length N as shown in Eq. (3).

$$h'(n) = \begin{cases} h(n)\ n = 0, 1 \ldots N-1, \\ 0, \quad otherwise. \end{cases} \tag{3}$$

An oscillatory pattern or ripples are observed in the magnitude response when the impulse response coefficients of the FIR filter are truncated. The number of ripples, both in the stopband and the passband are directly proportional to the increasing of the length of the FIR filter, while there is reduction in the width of the ripples [2]. The oscillatory ripples although being narrow, the height of the ripples are constant. The maximum ripples occur at the fiduciary edges or near the transition points. This undesirable trait is called as Gibb's phenomenon [3]. This effect appears as a fixed percentage overshoot and ripple before and after the discontinuity. This is true because it is impossible to obtain an infinite slope using only a finite number of terms. As stated, as the number of terms increase, the ripples do not decrease but are squeezed into a narrower interval about the discontinuity. Even in the infinite sum, this overshooting and undershooting persists and the complete series has flanges. Resulting, the series in trying to follow the discontinuity, overshoots the mark by about 18% [4] over a region before settling down to a correct value of unity. For this reason the rectangular window is not of much practical use and other window sequences w(n) alleviate this problem as seen in Fig. 1.

The optimal method represented by various algorithms such as Chebyshev approximation criterion [5, 6], Remez exchange algorithm [7], Park McClellan algorithm [8] have been developed and works on the concept of having equiripples in the pass-

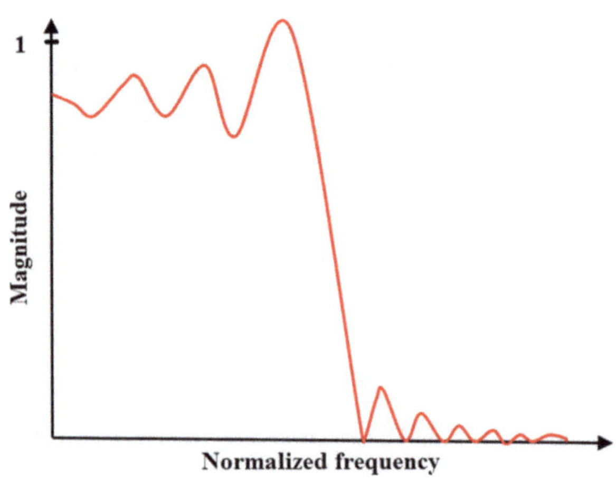

Fig. 1 Low pass filter designed with a rectangular window for a certain N

band and stopband. In the passband, the ripples oscillates between 1 − δp and 1 + δp and 0 and δs in the stopband. The main algorithm is an iterative process to determine the extremal frequencies of a filter whose amplitude frequency response satisfies the optimality condition. Algorithms such as Interpolated FIR filters [9], frequency response masking filters [10] among others FIR filter designs [11, 12] were reviewed before proposing a new technique using a linear phase low pass filter algorithm wherein the stopband, transition band and the passband regions of the filter magnitude response are modelled using trigonometric functions as it is obtained for the band pass filter design in [13]. In a similar way in [14], the sinusoidal trigonometric functions aid in computing the impulse response coefficients of the filters. In this method [14] the center frequency decides the band edges of the filter and the design parameters.

2 Proposed Methodology Using Linear Phase FIR Filter

In this section, the detailed design of the linear phase FIR low pass filter is proposed where, H (ω) is the magnitude of the filter response, δs is the stopband attenuation and δp is the passband ripple [15].

In the *passband region* of $0 \le \omega \le \omega_{cl}$, the frequency response is

$$H(\omega) = (1 - \delta_p) + \delta_p \cos(k_{pl}\,\omega) \tag{4}$$

$$\begin{aligned} &At\,\omega = 0; \quad H(0) = (1-\delta_p) + \delta_p = 1 \\ &At\,\omega = \omega_{cl};\; H(\omega_{cl}) = \ (1-\delta_p) + \delta_p \cos(k_{pl}\omega_{cl}) = 1 - \delta_p \\ &\therefore\ k_{pl} = \tfrac{\pi}{2\omega_{cl}} \end{aligned} \tag{5}$$

In the *sharp transition region* for $\omega_{cl} \le \omega \le \omega_{sl}$, the frequency response is,

$$H(\omega) = \delta_s + (1 - \delta_p - \delta_s)\cos k_{tl}(\omega - \omega_{cl}) \tag{6}$$

$$\begin{aligned} &At\,\omega = \omega_{cl};\; H(\omega_{cl}) = \delta_s + (1-\delta_p-\delta_s) = 1 - \delta_p \\ &At\,\omega = \omega_{sl};\; H(\omega_{sl}) = \delta_s + (1-\delta_p-\delta_s)\cos k_{tl}(\omega_{sl} - \omega_{cl}) = \delta_s \\ &\therefore\ k_{kl} = \tfrac{\pi}{2\,(\omega_{sl}-\omega_{cl})} \end{aligned} \tag{7}$$

In the *stop band region* for $\omega_{sl} \le \omega \le \pi$, the frequency response is,

$$H(\omega) = \delta_s - \delta_s \sin(k_{sl}(\omega - \omega_{sl})) \tag{8}$$

$$\begin{aligned} &At\,\omega = \omega_{sl};\; H(\omega_{sl}) = \delta_s - \delta_s \sin k_{sl}(\omega - \omega_{sl}) = \delta_s \\ &At\,\omega = \pi; \quad H(\pi) = \delta_s - \delta_s \sin k_{sl}(\pi - \omega_{sl}) = 0 \\ &\therefore\ k_{sl} = \tfrac{\pi}{2(\pi-\omega_{sl})} \end{aligned} \tag{9}$$

Passband (k_{pl}), transition (k_{tl}) and stopband (k_{sl}) are filter design parameters derived in Eqs. (5), (7) and (9) respectively.

2.1 Expressions for Impulse Response Coefficients for the FIR LPF

The impulse response coefficients h (n) for the FIR LPF are obtained by computing the integral limits of the three regions of the filter magnitude response.

$$h(n) = \frac{1}{\pi}\left\{\int_{0}^{\omega_{cl}} H(\omega)\cos k\,\omega\,\partial\omega + \int_{\omega_{cl}}^{\omega_{sl}} H(\omega)\cos k\,\omega\,\partial\omega + \int_{\omega_{sl}}^{\pi} H(\omega)\cos k\,\omega\,\partial\omega\right\} \tag{10}$$

Solving the 1st term from Eq. (10)

$$\text{1st term} = h_{1l}(n) = \frac{1}{\pi}\int_{0}^{\omega_{cl}} \left[\left(1-\delta_p\right) + \delta_p \cos k_{pl}\omega\right]\cos k\,\omega\,\partial\omega$$

$$= \frac{\left(1-\delta_p\right)\sin k\omega_{cl}}{k\pi} + \frac{\delta_p}{\pi\left(k_{pl^2} - k^2\right)}\left[k_{pl}\sin\left(k_p r\omega_{cl}\right)\cos(k\omega_{cl}) - k\cos\left(k_{pl}\omega_{cl}\right)\sin(k\omega_{cl})\right] \tag{11}$$

Solving the 2nd term from Eq. (10)

$$\text{2nd term} = h_2(n) = \frac{1}{\pi}\int_{\omega_{cl}}^{\omega_{sl}} \left[\delta_s + \left(1-\delta_p-\delta_s\right)\cos\, k_{tl}(\omega-\omega_{cl})\right]\cos\, k\omega\,\partial\omega$$

$$= \frac{\delta_s}{k\pi}[\sin k\omega_{sl} - \sin k\omega_{cl}] + \frac{\left(1-\delta_p-\delta_s\right)}{\pi\left(ktl^2 - k^2\right)}\begin{bmatrix} k_{tl}\sin k_{tl}(\omega_{sl}-\omega_{cl})\cos(k\omega_{sl}) - k\cos k_{tl}(\omega_{sl}-\omega_{cl})\sin(k\omega_{sl}) \\ +k\sin(k\omega_{cl}) \end{bmatrix} \tag{12}$$

Solving the 3rd term from Eq. (10)

$$\text{3rd term} = h_3(n) = \frac{1}{\pi}\int_{\omega_{sl}}^{\pi} [\delta_s - \delta_s\sin(ksl(\omega-\omega_{sl}))]\cos k\omega\,\partial\omega$$

$$= \frac{\delta_s}{k\pi}[\sin k\,\pi - \sin k\omega_{sl}] + \frac{\delta_s}{\pi\left(k_{sl}^2 - k^2\right)}$$

$$[k_{sl}\cos[(k_{sl}(\pi-\omega_{sl})]\cos k\,\pi + k\sin[(k_{sl}(\pi-\omega_{sl})]\sin k\,\pi - k_{sl}\cos(k\omega_{sl})] \tag{13}$$

By substituting Eqs. (11)–(13) in Eq. (10) we obtain the expression for the low pass filter impulse response h (n). Where $k \neq k_{tl}$, k_{pl} *and* k_{sl}.

$$h(n) = \left\{ \left(\frac{\delta_s}{k\pi}\right) \left[\sin(k\omega_{sl}) - \sin(k\omega_{cl}) + \sin(k\pi) - \sin(k\omega_{sl})\right] + \frac{(1-\delta_p)\sin(k\omega_{cl})}{k\pi} \right\}$$
$$+ \left\{ \left(\frac{\delta_p}{\pi\left(k_{pl}^2 - k^2\right)}\right) \left[k_{pl}\sin(k_p r\omega_{cl})\cos(k\omega_{cl}) - k\cos\left(k_{pl}\omega_{cl}\right)\sin(k\omega_{cl})\right] \right\}$$
$$+ \left\{ \left(\frac{(1-\delta_p-\delta_s)}{\pi\left(ktl^2 - k^2\right)}\right) \left[k\sin(k\omega_{cl}) + k_{tl}\sin(k_{tl}(\omega_{sl}-\omega_{cl}))\cos(k\omega_{sl}) - k\cos(ktt(\omega_{sl}-\omega_{cl}))\sin(k\omega_{sl})\right] \right\}$$
$$+ \left\{ \left(\frac{\delta_s}{\pi\left(k_{sl}^2 - k^2\right)}\right) \left[k_{sl}\cos(k_{sl}(\pi-\omega_{sl}))\cos(k\pi) + k\sin(k_{sl}(\pi-\omega_{sl}))\sin(k\pi) - k_{sl}\cos(k\omega_{sl})\right] \right\} \quad (14)$$

Equation (14) is the expression for the LPF model impulse response h(n).

2.2 *Expression for the Frequency Response of the FIR Low Pass Filter*

We selected the symmetric impulse response, h(n) = h(N − 1 − n) for N Even and the appropriate type of frequency response for the linear phase FIR LPF as shown below [1, 4].

$$H_r(\omega) = 2\sum_{n=0}^{\left(\frac{N}{2}\right)-1} h(n)\cos\left(\omega\left(\frac{N-1}{2} - n\right)\right) \quad (15)$$

This filter design is most suitable for LPFs as the H(0) gives a maximum value, while H(π) = 0.

3 Results

The performance of the FIR low pass filter for various filter orders (N) are clearly depicted in Fig. 2. Where in the passband loss is as low as 6.34% for N = 200 and further reduces to nearly 2% at larger filter order (N = 2000). There is a reduction of Gibb's phenomenon with our proposed linear phase FIR LPF designs. As we know the conventional FIR filters, the peak passband ripple due to Gibb's phenomenon is about 18%, our passband loss is very much low as compared to the conventional peak passband ripple value. Using our proposed low pass FIR filters, we also observed from Figs. 3 and 4, that (i) the passband losses are quite low, (ii) The ripple decreases for higher filter order and (iii) The filter exhibited sharp transition region as low of 2π. The magnitude response for various filter orders ranging from N = 200 to N =

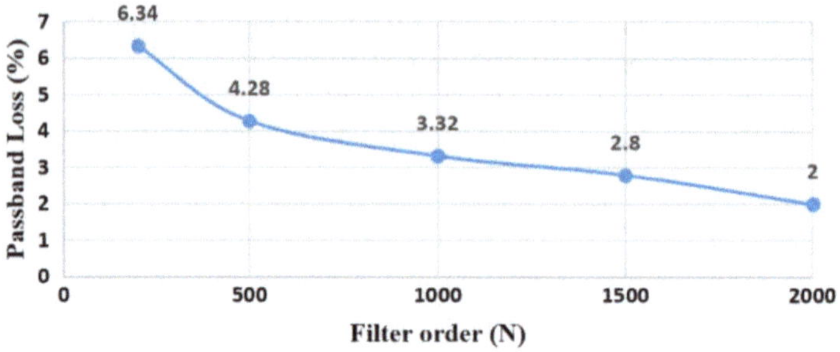

Fig. 2 Passband loss of the LPF design for various filter order

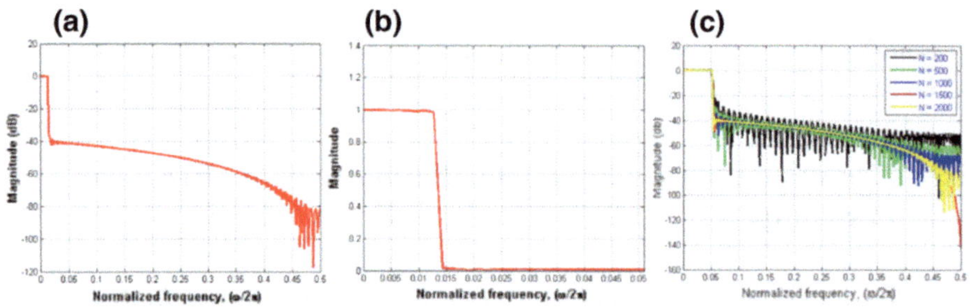

Fig. 3 Linear phase FIR LPF **a** magnitude response (N = 1000) **b** linear plot (N = 1000) **c** magnitude response for various filter order

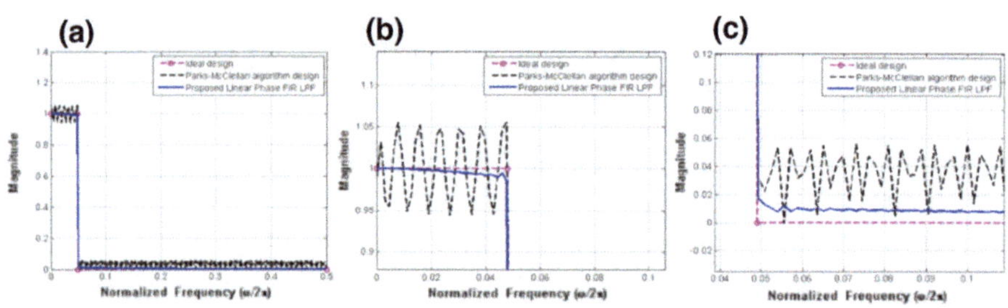

Fig. 4 Magnitude response of PM algorithm as compared with the proposed LPF **a** linear plot **b** magnified pass band view **c** magnified stopband view

2000 were also plotted to evaluate the performance of the proposed LPF as shown in Fig. 3c. These filters are unlike the classical filters, possess a narrow stopband and/or passband. The stopband attenuation was recorded to be 40 dB. The proposed design was compared with the optimal linear phase FIR method such as the Parks McClellan (PM) algorithm. The PM algorithm exhibits large ripple in the pass and the stop bands as compared to our proposed algorithm as shown in Fig. 4a–c.

The proposed linear phase FIR LPF was used to filter an electroencephalogram (EEG) signal whose recording was taken for a single subject (activity: relaxed and alert) for a duration of 5 min. As many as 20 EEG channels were collected with

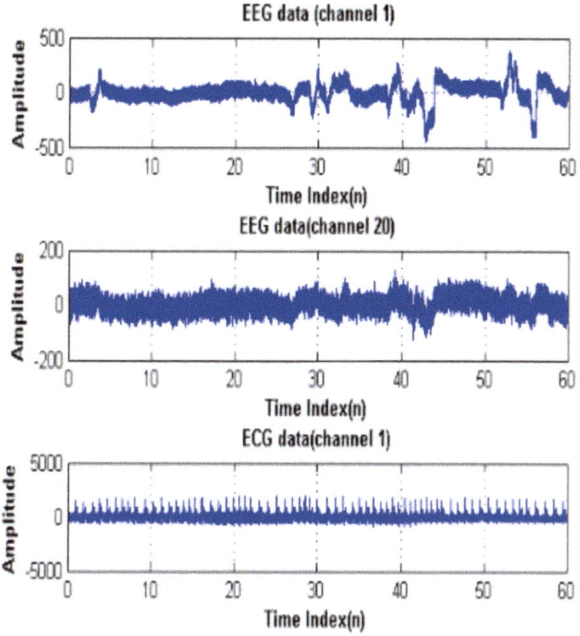

Fig. 5 Real time EEG signals from a healthy subject

the sampling frequency (fs) of 256 Hz. To enable and filter in the Theta–Alpha frequencies from the raw EEG signal [16–18], the sharp transition capable linear phase FIR LPF set its passband edge to 12 Hz and the stopband edge at 13 Hz. Figure 5 shows the extracted real time EEG signal from the subject from channel 1 and 20 (other channel signals are not shown here). Figure 6a displays the FFT spectrum of the original EEG (in blue) and the FFT of the filtered EEG after using our proposed FIR filter (in red). The high frequency artifacts, frequencies above

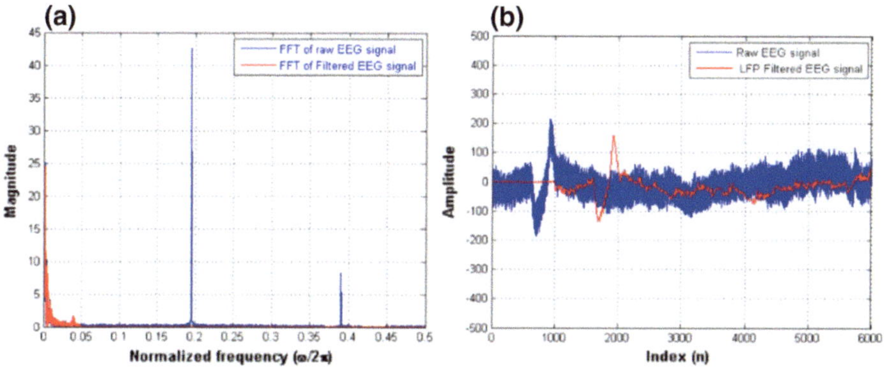

Fig. 6 Real time EEG signal along with filtered EEG signal **a** frequency domain **b** time domain

13 and 50 Hz power line interference (PLI) and its harmonics are also filtered and suppressed. The time domain traces in Fig. 6b displays the original EEG signal (in blue) and the noise free EEG signal (shown in red) which can be further analyzed and processed.

4 Conclusions

Using the trigonometric functions of frequency the impulse response coefficients are computed for the three regions of the filter response in the frequency domain. The proposed filter model stressed on achieving a sharp transition and a flat passband. As the filter gets sharper in the transition region, more oscillations or ripples will be the frequency response near the edge of the passband, a trait described as Gibb's phenomenon. However our proposed filter model achieved a fair trade-off between the transition bandwidth and the Gibb's phenomenon. Using our methodology, a low passband ripple of 2% was achieved for filter order 2000 with a stopband attenuation of 40 dB. It was also seen that our proposed method had a flat passband and stopband as compared to the large ripples seen in the passband and stopband bands using the Parks McClellan algorithm Thus a threefold satisfactory performance was achieved in all the three bands namely passband, transition and stopband. Our proposed linear phase FIR LPF was effectively used to filter out PLI and higher unwanted high frequencies from the real time EEG signals with sharp transition filter band edges.

Acknowledgements The authors would like to thank Mr. Noel Tavares, research scholar from the Electronics department, Goa University for collecting the EEG clinical data for testing our filter algorithm.

References

1. Ifeachor EC, Jervis BW (2002) Digital signal processing: a practical approach. Pearson Education, pp 367–379
2. Proakis JG, Manolakis DG (1992) Digital signal processing-principles, algorithms and applications
3. Gibbs AJ (1970) The design of digital filters in Australian. Telecommun Res J 4:29–34
4. Johnson JR (1989) Introduction to digital signal processing. Prentice Hall, Englewood Cliffs, NJ, pp 426-440
5. Rabiner LR et al (1975) FIR digital filter design techniques using weighted Chebyshev approximation. Proc IEEE 63(4):595–610
6. Alkhairy A et al (1991) Design of FIR filters by complex Chebyshev approximation. In: IEEE international conference on acoustics, speech, and signal processing, ICASSP-91, pp 1985–1988
7. Saramäki T, Lim YC (2003) Use of the Remez algorithm for designing FRM based FIR filters. Circ Syst Signal Process 22(2):77–97
8. Vaidyanathan P (1985) Optimal design of linear phase FIR digital filters with very flat passbands and equiripple stopbands. IEEE Trans Circuits Syst 32(9):904–917

9. Saramaki T et al (1988) Design of computationally efficient interpolated FIR filters. IEEE Trans Circ Syst 35(1):70–88
10. Lim Y (1986) Frequency-response masking approach for the synthesis of sharp linear phase digital filters. IEEE Trans Circ Syst 33(4):357–364
11. Jing Z, Fam A (1984) A new structure for narrow transition band, low pass digital filter design. IEEE Trans Acoust Speech Signal Process 32(2):362–370
12. Henzel N, Leski JM (2013) Generalized constraint design of linear-phase FIR digital filters. In: Proceedings of the 8th international conference on computer recognition systems CORES 2013. Springer, Heidelberg, pp 53–62
13. Marchon N et al (2018) Linear phase sharp transition BPF to detect noninvasive maternal and fetal heart rate. J Healthc Eng Hindawi 2018:1–14
14. Rodrigues J, Pai KR (2005) New approach to the synthesis of sharp transition FIR digital filter. In: Proceedings of the IEEE international symposium industrial electronics, vol 3, pp 1171–1173
15. Marchon N et al (2018) Monitoring of fetal heart rate using sharp transition FIR filter. Biomed Signal Process Control 44:191–199
16. Britton JW et al (2016) Electroencephalography (EEG): an introductory text and atlas of normal and abnormal findings in adults, children, and infants. American Epilepsy Society, Chicago
17. Akiyama M et al (2017) Theta-alpha EEG phase distributions in the frontal area for dissociation of visual and auditory working memory. Sci Rep 7:42776
18. Malmivuo J, Plonsey R (1995) Bioelectromagnetism: principles and applications of bioelectric and biomagnetic fields. Oxford University Press, USA

PLS-Based Multivariate Statistical Approach for Soft Sensor Development in WWTP

Barasha Mali and S. H. Laskar

Abstract Multivariate Statistical Process Control is a projection method of projecting a high- dimensional model space with a number of measured variables to a low-dimensional space. The different methods include Principal Component Analysis, Principal Component Regression, Partial Least Squares Regression, Finite Impulse response, Autoregressive exogenous input and autoregressive moving average, etc. The advantage of using Multivariate Statistical Process Control is that it identifies low-dimensional quality process data while reducing the variability in the process and increasing the product quality. The paper aims to find the low- dimensional information-rich space for soft sensor design using Partial Least Squares-based Multivariate Statistical Process Control technique for the controlling variables in a Wastewater Treatment Plant. The input considered here is the stored data of the actual process variables obtained from the plant. This is carried out using the 14 days data for the three weather conditions, dry rain and storm available from the benchmark model. The performance of the applied method is verified using scatter plot and R-squared.

Keywords Multivariate process control · Soft sensor · Wastewater treatment plant · Principal component analysis · Partial least squares · Regression

1 Introduction

Process Industries are mainly concerned about the three issues relating to their product. First is the quality of the product or the specifications according to the customer [20]. Also, analysis of the product before the production is more economical as

B. Mali (✉) · S. H. Laskar
National Institute of Technology Silchar, Cachar 788010, Assam, India
e-mail: barashamali@gmail.com

S. H. Laskar
Sant Longowal Institute of Engineering and Technology,
Longowal, Sangrur 148106, Punjab, India

C. Shreesha and R. D. Gudi (eds.), *Control Instrumentation Systems*,
Lecture Notes in Electrical Engineering 581,
https://doi.org/10.1007/978-981-13-9419-5_11

compared to the analysis after the production. Second, the delivered product should be error free, which depends on the different materials, equipment, methods and reference instructions during the process.

According to [17] based on the materials and methods used, the time required and the final produced quality demand for a particular product may change and accordingly the price becomes the third issue that has to be considered. These motives make the system complex and also increases the number of variables involved in the process. Therefore, univariate statistical analysis in the traditional Statistical Process Control like mean, median, standard deviation, etc. are replaced by an advanced statistical approach known as the Multivariate Statistical Process Control (MSPC). This type of analysis identifies the different crucial variables, the relations between them and the underlying patterns in the input data set. These relationships help us to understand the process deeper and help to decide the best strategy by predicting the response of different control strategies before implementing them practically as mentioned by [25].

MPSC is seen to be used in different industries like chemical industries as in [10], food industries as in [4] and [2], paper and pulp industries as in as in [24], water and wastewater industries as in [24] and [7], pharmaceutical industries as in [22] and so on. Historic data set is acquired from a plant in different available conditions to obtain the relation between the available parameters. They are also studied to extract the essential features and the virtual parameter indicating the information about deviation from some controlled range is obtained using various techniques from the multivariate data sets.

Different approaches of Multivariate Statistical Process Control (MSPC) such as Principal Component Analysis or PCA, Independent Component Analysis or ICA, Partial Least Squares or PLS, Partial Least Squares Regression or PLSR, Support Vector Regression or SVR, Moving Average or MA, Autoregressive or AR, Autoregressive Moving Average or ARMA, etc. are discussed in [11, 13, 14, 19]. In this paper Partial Least Squares Regression (PLSR) method is discussed and used to obtain lesser number of process variables as compared to the actual variables obtained from a Wastewater Treatment Plant. The reduced data set is used to identify the dry, rain and storm datasets and is compared with the most frequently used Principal Component Regression (PCR). These approaches can be used for soft sensor development of the plant to obtain proper control without delay in the absence of hardware sensor.

The paper is organized as follows: Sect. 2 discusses the MPSC techniques in industrial process monitoring and control as seen in various papers in the last 3 years, Sect. 3 discusses the Wastewater Treatment data and the MPSC methods (PLSR) used in this paper. Comparison of both the methods in terms of R-squared parameter and plots comparing the original and predicted flow is also included in this section which is the main contribution of the paper, Sect. 4 analyses the compared results obtained and Sect. 5 concludes the paper giving ideas about the future work.

2 MPSC Techniques in Literature

MPSC is popular in the field of chemometrics and is now been used in various fields as can be seen in Table 1. It shows the different proposed techniques during the last three years in chemical industries food industries, pharmaceutical industries including large-scale industries.

Table 1 MPSC related works in the last three years

Papers	MPSC-related work
Cheng et al. [5]	Decision tree (DT)-based ensemble classifier is used to identify out of control data
Ji et al. [11]	Incipient fault detection using generic fault detection index is proposed
Sikder et al. [21]	Mahalanobis distance, Taguchis orthogonal array, and the main effect plot are used to identify the variables influencing the erroneous situations
Silva et al. [22]	PCA used for continuous tablet (ConsiGma-25) manufacturing line monitoring
Silva et al. [23]	Orthogonal PLS (OPLS) is proposed for online monitored cocrystallization process with near-IR spectroscopy
Cohen et al. [6]	DeWave control chart for monitoring process variability is proposed
Lim et al. [15]	Willingness to use SPC in food industry is discussed
Ogden et al. [18]	Uses MPSC in mechanical systems for automating and improving anomaly detection
Zhang [28]	Two methods using K-Nearest Neighbour (KNN)and support vector machine (SVM) in control and out of control data are discussed to check the new parameters variability
Yin et al. [27]	Data-based technique for modern industries is discussed here
Addeh et al. [1]	Control Chart Recognition Pattern (CCP) design is investigated in the feature extraction and classifier module
Fan et al. [8]	Filtering kernel independent component analysis principal component analysis (FKICAPCA) is validated in a Tennessee Eastman (TE) process
Grassi et al. [9]	Capability of a beer fermentation is accessed using Principal component analysis (PCA), partial least squares (PLS) and locally weighted regression (LWR)
Jiang et al. [12]	Just-in-time reorganized PCA model and conventional PCA models are compared for chemical processes
Yin et al. [26]	The basic data-driven methods for industrial processes in large scale is reviewed
Zhao and Gao [29]	Fault-relevant Principal Component Analysis (FPCA) is proposed and simulated for Tennessee Eastman (TE) process

3 WWTP Data Analysis Using PLSR

Statistical methods in data analysis is an important tool in chemometrics and therefore is used to analyse the relation between the various data sets obtained from a WWTP. We have three data matrices with dry weather data, rain weather data and storm weather data. The data set consists of 14 days data and 13 process variables with one input variable.

Reduced dimensional parameters obtained using MPSC techniques like PCA and PLS are compared with the traditional SPC charts in [3, 16]. The measurement of process deviation from the desired point can be done using multivariate methods where all the process variables are treated simultaneously unlike the traditional methods.

The classifier algorithm separates the in-control and out-control product variables to identify the fault based on multivariate χ^2 and $T2$ charts in traditional methods, which sometimes fail when measured variables for the plant are highly correlated and the covariance matrix becomes singular. Therefore, they are unable to do the same task and a modified technique is needed to overcome such difficulty.

Principal Component analysis, which reduces the high-dimensional data to a low-dimensional data based on the variance or Partial least squares using latent variables concept can be used for this purpose of handling the difficulty as seen in using the traditional methods. Also for any practical situation, the data is almost always highly correlated and noisy and charts cannot be relied. Therefore, for practical cases, PCR and PLSR are used.

Methods like Neural Network and Fuzzy Logic can also be used to modify the available data. However, the paper contribution is limited to variable reduction for the above-mentioned WWTP using PLSR technique.

PLSR is a modified PCR where instead of finding the scores and loadings from a single data matrix, two matrices containing input and output are used to find the scores and loadings with the assumption that the output variable is dependent on the input variables. So after obtaining the model, the output can be predicted only by using the model and the input data sets that make up the input data matrix.

The basic equation relating the n samples with p predictor variables and r response variables for PCR and PLSR is given by Eq. 1

$$Y = XB + E \tag{1}$$

where, X is a $n \times p$ predictor matrix, Y is a $n \times r$ response matrix, B is a $p \times r$ regression coefficient matrix and E is a $n \times r$ noise matrix.

To avoid the correlation among the factor scores of PCR and PLSR, the factor scores are produced as linear combination of original n predictor variables. A linear regression model as in Eq. 2 is considered where T is the score factor matrix and equal to XW with Weight matrix W, Q is the matrix of the loadings or regression coefficients of T and other parameters remain the same.

$$Y = TQ + E \tag{2}$$

Once Q is known, the model developed becomes equivalent to Eq. 1 and can be used as a predictive model with $B = WQ$.

This type of prediction models can be used to replace hardware sensors when they are not available or when there is some error in it. The difference between the two methods is in determining the factor score matrix $T = BW$. In PCR, the W weight matrix is obtained by considering only the correlation between the predictor variables while in PLSR, weight matrix W is obtained considering the predictor variables as well as the response variables.

4 Results and Analysis

More variables doesn't always give good PLSR model for soft sensor development. So to find the minimum number of components that will give the best fit model for soft sensor development can be found from the percentage of variance explained in the response variable versus the number of principal components plot. Figure 1 shows the plot of the number of principal components in the X-axis and the percentage of variance of the response variable in the Y-axis and it is seen that the number of two components gives the best fit for the PLSR method for the wastewater treatment plant considered in this paper.

Actual Response versus Predicted Response scatter plots help to visualize and analyse data in a simple yet clear way. The closeness of the modified data points to the regressed diagonal indicates the performance of the predicted results. Figure 2 shows the Fitted Response versus Observed Response comparison plots for dry weather data considering one Principal Component, two Principal Components and three Principal Component. The two Principal Component plot lies close to the regressed diagonal line and hence can be concluded as the method which can be used to develop a soft sensor for the considered wastewater treatment plant. Similarly, the Fitted Response versus Observed Response comparison plots for rain weather

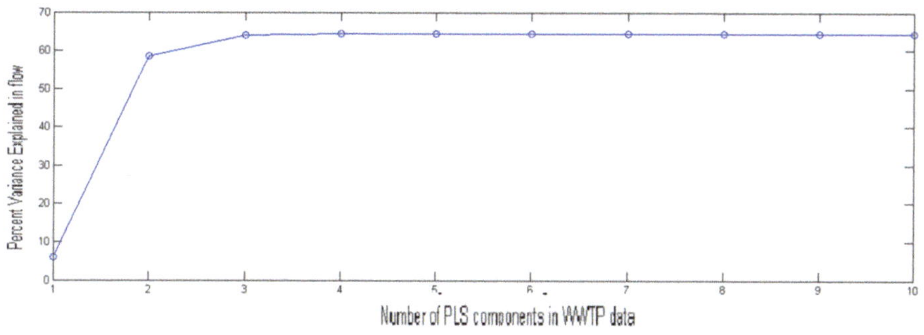

Fig. 1 Percentage variance in the response variable versus the number of principal components plot

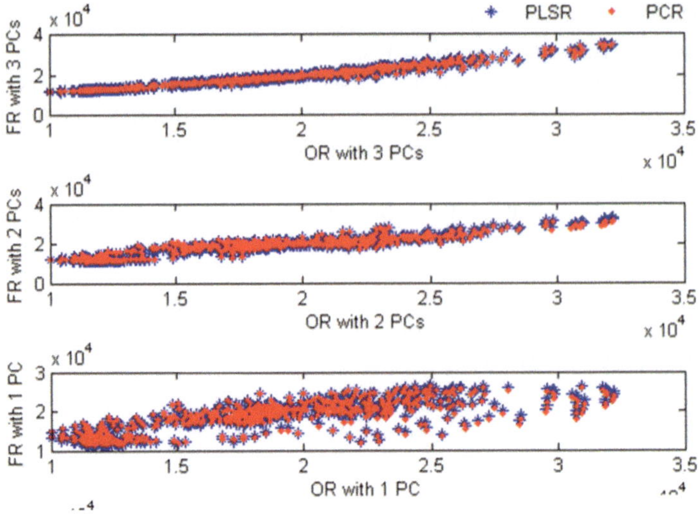

Fig. 2 Fitted Response versus Observed Response for dry weather data **a** with 1 PC, 2 PCs and 3 PCs

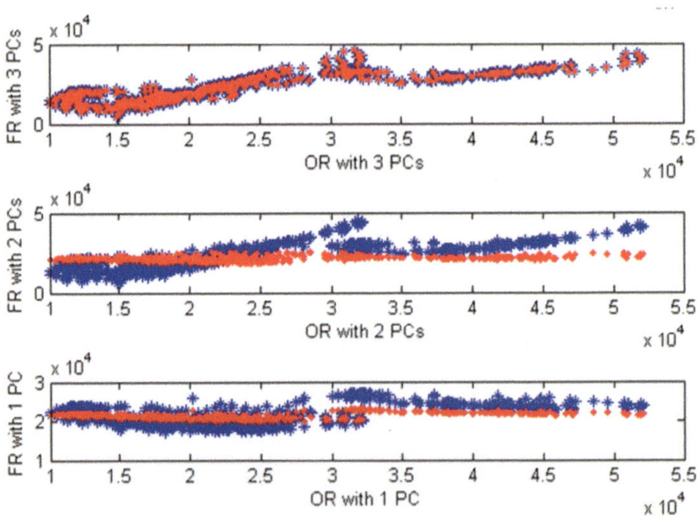

Fig. 3 Fitted Response versus Observed Response for rain weather data **a** with 1 PC, 2 PCs and 3 PCs

data and storm weather data considering one Principal Component, two Principal Components and three Principal Component is shown in Figs. 3 and 4 and the plot considering two principal components is close to the diagonal in both the cases.

Another measure of fit performance is the R-squared value. The model with high R-squared value has a better fit and is close to the diagonal and the model with lower

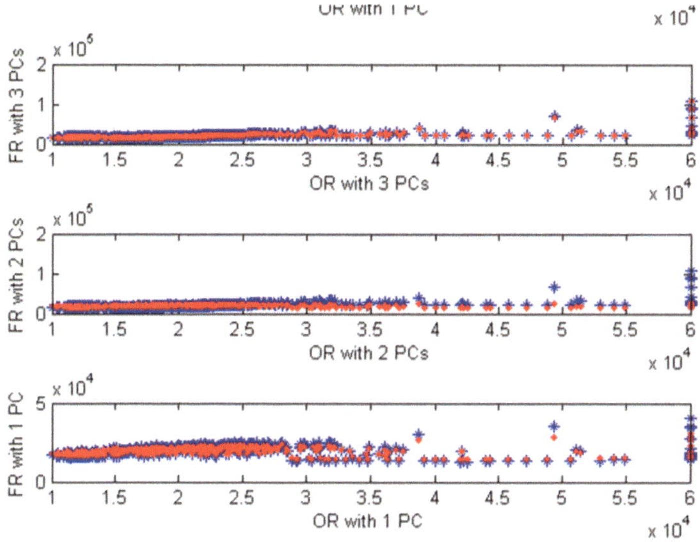

Fig. 4 Fitted Response versus Observed Response for storm weather data **a** with 1 PC, 2 PCs and 3 PCs

Table 2 Rsquare compared for PCR and PLSR with increasing number of PCs

	1 PC	2 PCs	3 PCs	4 PCs
Dry weather data				
PCR	0.5871	0.8547	0.9625	0.9629
PLSR	0.6410	0.8916	0.9626	0.9629
Rain weather data				
PCR	0.0061	0.0204	0.6396	0.6442
PLSR	0.0601	0.4736	0.6400	0.6442
Storm weather data				
PCR	0.0665	0.0687	0.4725	0.5123
PLSR	0.1209	0.4736	0.4756	0.5123

R-squared value shows that the model results in data points with dispersed points across the diagonal. The R-squared values of the models from the three data sets dry weather, rain weather and storm weather is shown in Table 2.

5 Conclusion and Discussion

PLS-based MPSC therefore reduces the 13 process variables and 1 input variable to 2 or 3 Principal Component to give the same information. Here, the PCs help to identify the type of data input to the wastewater treatment plant which is very unpredictive otherwise. Such early detection of the type of input enables a plant operator to take necessary action to maintain the type of input. Overall equipment efficiency is also increased as it reduces the maintenance time by implementing corrective and preventive actions before the drift in the process parameters becomes unfavourable. Dry weather data set is considered as a reference against which the rain weather and storm weather data will be compared to differentiate the unfavourable situations. Again, parameters like R-squared value also helps to detect the model fit and hence indicate how correct the predicted data is. This method which is applied in three data sets of a wastewater treatment plant data set may be verified with a large data set to obtain better predictions.

References

1. Addeh J, Ebrahimzadeh A, Azarbad M, Ranaee V (2014) Statistical process control using optimized neural networks: a case study. ISA Trans 53(5):1489–1499
2. Asadi A, Zinatizadeh AA, Van Loosdrecht M (2016) A novel continuous feed and intermittent discharge airlift bioreactor (CFIDAB) for enhanced simultaneous removal of carbon and nutrients from soft drink industrial wastewater. Chem Eng J 292:13–27
3. Bersimis S, Panaretos J, Psarakis S (2005) Multivariate statistical process control charts and the problem of interpretation: a short overview and some applications in industry. In: Proceedings of the 7th hellenic European conference on computer mathematics and its applications. Athens, Greece
4. Cantero-Tubilla B, Cantero DA, Martinez CM, Tester JW, Walker LP, Posmanik R (2017) Characterization of the solid products from hydrothermal liquefaction of waste feedstocks from food and agricultural industries. J Supercrit Fluids
5. Cheng CS, Lee HT (2017) Identifying the source of variance shifts in multivariate statistical process control using ensemble classifiers. In: Theory and practice of quality and reliability engineering in Asia industry. Springer, Singapore, pp 27–39
6. Cohen A, Tiplica T, Kobi A (2016) Design of experiments and statistical process control using wavelets analysis. Control Eng Pract 49:129–138
7. Dhiman N, Shukla SP, Kisku GC (2017) Statistical optimization of process parameters for removal of dyes from wastewater on chitosan cenospheres nanocomposite using response surface methodology. J Clean Prod 149:597–606
8. Fan J, Qin SJ, Wang Y (2014) Online monitoring of nonlinear multivariate industrial processes using filtering KICAPCA. Control Eng Pract 22:205–216
9. Grassi S, Amigo JM, Lyndgaard CB, Foschino R, Casiraghi E (2014) Beer fermentation: monitoring of process parameters by FT-NIR and multivariate data analysis. Food Chem 155:279–286
10. Hao J, Liebeke M, Sommer U, Viant MR, Bundy JG, Ebbels TM (2016) Statistical correlations between NMR spectroscopy and direct infusion FT-ICR mass spectrometry aid annotation of unknowns in metabolomics. Anal Chem 88(5):2583–2589
11. Ji H, He X, Shang J, Zhou D (2017) Incipient fault detection with smoothing techniques in statistical process monitoring. Control Eng Pract 62:11–21
12. Jiang Q, Yan X (2014) Just in time reorganized PCA integrated with SVDD for chemical process monitoring. AIChE J 60(3):949–965
13. Kruger U, Zhou Y, Irwin GW (2004) Improved principal component monitoring of large-scale processes. J Process Control 14(8):879–888

14. Lieftucht D, Kruger U, Irwin GW (2005, December) Compensation terms to improve fault detection in multivariate auto-correlated processes. In: 44th IEEE conference on decision and control, 2005 and 2005 European control conference. CDC-ECC'05. IEEE, pp 3827–3831
15. Lim SAH, Antony J (2016) Statistical process control readiness in the food industry: development of a self-assessment tool. Trends Food Sci Technol 58:133–139
16. MacGregor JF, Kourti T (1995) Statistical process control of multivariate processes. Control Eng Pract 3(3):403–414
17. Oakland JS (2007) Statistical process control. Taylor & Francis, Routledge
18. Ogden D, Arnold T, Downing W (2016, October) A multivariate statistical approach for improved automated process control and anomaly detection in mechanical systems. In: 2016 IEEE international symposium on systems engineering (ISSE). IEEE, pp 1–7
19. Papazoglou M (1998) Multivariate statistical process control of chemical processes, Doctoral dissertation, Newcastle University
20. Shang L, Liu J, Zhang Y (2016) Recursive fault detection and identification for time-varying processes. Ind Eng Chem Res 55(46):12149–12160
21. Sikder S, Sikder S, Panja SC, Panja SC, Mukherjee I, Mukherjee I (2017) An integrated approach for multivariate statistical process control using Mahalanobis-Taguchi system and Andrews function. Int J Qual Reliab Manag 34(8):1186–1208
22. Silva AF, Sarragua MC, Fonteyne M, Vercruysse J, De Leersnyder F, Vanhoorne V, De Beer T (2017) Multivariate statistical process control of a continuous pharmaceutical twin-screw granulation and fluid bed drying process. Int J Pharm
23. Silva AF, Sarragua M C, Ribeiro PR, Santos AO, De Beer T, Lopes JA (2017) Statistical process control of cocrystallization processes: a comparison between OPLS and PLS. Int J Pharm 520(1):29–38
24. Sridhar R, Sivakumar V, Thirugnanasambandham K (2016) Response surface modeling and optimization of upflow anaerobic sludge blanket reactor process parameters for the treatment of bagasse based pulp and paper industry wastewater. Desalination Water Treat 57(10):4345–4356
25. Westad F (2012) Monitoring chemical processes for early fault detection using multivariate data analysis methods. Whitepaper CAMO software
26. Yin S, Ding SX, Xie X, Luo H (2014) A review on basic data-driven approaches for industrial process monitoring. IEEE Trans Ind Electron 61(11):6418–6428
27. Yin S, Li X, Gao H, Kaynak O (2015) Data-based techniques focused on modern industry: an overview. IEEE Trans Ind Electron 62(1):657–667
28. Zhang C (2016) Statistical process control for multivariate data and lifetime data. Doctoral dissertation
29. Zhao C, Gao F (2014) Fault-relevant principal component analysis (FPCA) method for multivariate statistical modeling and process monitoring. Chemom Intell Lab Syst 133:1–16

PLX-DAQ-Based Wireless Battery Monitoring System for Obstacle Avoidance Robot

M. V. Sreenivas Rao and M. Shivakumar

Abstract Most mobile robotic systems draw power from batteries which have a limited power life. Monitoring the status of the battery power on the robot is therefore important for autonomous robotic systems. The wireless system for monitoring lead–acid battery of obstacle avoidance robot has been developed. The system employs sensors, microcontroller ATMEGA 328, and Bluetooth module. The parameters such as voltage and current of battery are checked by the battery monitoring system during discharging of the battery. The varying discharging voltage and current values of the battery are detected using voltage divider circuit and Hall Effect current sensor, respectively. Battery voltage and current data will be transferred in real time to the microcontroller, and then it will be transmitted to display device using Bluetooth communication. In this work, monitored battery data will be displayed on Microsoft Excel of laptop Personal Computer (PC) with Parallax Data Acquisition tool (PLX-DAQ). The real-time data of voltage and current will be indicated by the system in the tabulated form along with graphical display.

Keywords Battery monitoring · PLX-DAQ · Obstacle avoidance robot · Bluetooth module

1 Introduction

Lead–acid batteries have numerous applications in many household and industrial appliances as their power source. The greater number of mobile robots depends on battery power for their operation. Batteries used in mobile robot require constant monitoring to ensure uninterrupted power supply for efficient operation. The mobile robot carrying battery monitoring system regularly checks the key operational param-

M. V. Sreenivas Rao (✉) · M. Shivakumar
Department of EIE, GSSSIETW, Mysuru, India
e-mail: mvsrao06@yahoo.com

M. Shivakumar
e-mail: drshivakumarm@gsss.edu.in

C. Shreesha and R. D. Gudi (eds.), *Control Instrumentation Systems*,
Lecture Notes in Electrical Engineering 581,
https://doi.org/10.1007/978-981-13-9419-5_12

eters such as voltage and current during discharging process. In this work, we have employed a wireless data acquisition system to monitor voltage and current output of the lead–acid battery of obstacle avoidance robot. The battery monitoring system used for acquiring voltage and current of lead–acid battery during discharging process is one of the main components of obstacle avoidance robot.

1.1 Objective/Problem Statement

In the real social environment, the mobile robot should operate with greater autonomy without human intervention. Autonomous mobile robotic systems consume power from batteries, which have shorter power life. This causes a greater challenge for autonomous mobile robot. The status of battery power monitoring in robotic system plays a significant role to check its power requirements while performing operations. If the battery voltage gets depleted and reaches the threshold level, the robot needs to locate the charging station to get recharged. This paper describes the specially developed wireless battery monitoring system for obstacle avoidance robot. The problem mentioned can be dealt with by continuously measuring discharging voltage and current values of lead–acid battery of obstacle avoidance robot.

2 Related Work

In recent years, different methods for wireless battery monitoring systems have been proposed. Oka Danil Saputra et al. have designed a system, which acquires data of lead–acid battery from a remote location for electric forklift. WLAN is used as the communication network for the data transmission. The system measures the current, voltage, SOC, FCC, and battery remaining capacity. The measured data will be transferred to the server by using WLAN technology. At the duration of one-minute interval of time, the data is received regularly. Graphical user interface is used to provide the battery parameters output data in the form of table format. For the data communication, WLAN core network is used for data communication between device and server [1]. Ashish Runiyar et al. have proposed multiple lead–acid batteries monitoring using IOT. The parameters of interest are voltage, current, SOC, battery's acid level, and the remaining charge capacity monitored. For the data transmission, wireless local area network used for collecting information related to all the batteries connected is analyzed on a personal computer. The data acquired from multiple batteries connected is studied by using a control protocol. The Android device is used to display the acquired data and stored in MySQL server database [2]. Vaibhav Verma et al. have presented a battery monitoring system by making use of NI input analog modules with National Instruments LabVIEW. The front panel displays the instantaneous data of each parameter with a graphical profile plotted continuously. The battery performance is tested at different temperatures. The real-time current and voltage plots for battery discharge profile are obtained precisely on the front panel of virtual instruments, respectively. It was shown that the battery after discharging at a higher temperature greater than 50 °C, self-discharging starts [3]. Chi Yuan Lee et al. have demonstrated that the overcharging of lithium battery may affect the voltage and current thereby producing battery instability. As a safety precaution, the information about the battery internal status condition will be reported in advance. The three-in-one microsensor was embedded in a coin-like cell. The batteries performance can be monitored by the single-integrated sensor, which can measure the internal temperature, current, and voltage instantly without deviating in the operation of the lithium battery [4]. Anif Jamaluddin et al. have designed a battery monitoring system using wireless method for electric vehicle. The features such as current, voltage, and temperature values are sent using short-range communication protocol. A LabVIEW program is used to display the battery parameters on personal computer and parameters on smartphone. The software made use is a custom-based program that executes the communication protocol. It was written in Java using X code IDE provided by the Apple Inc. The X code includes programming, debugging, compiling, and the simulation of the code and consist of built-in objects and libraries, which helps in the development process [5]. Tadej Tasner et al. have proposed easy-to-use platform for measurements based on Bluetooth communication with the smart devices. The wireless link of sensors is accomplished by implementing them by Bluetooth device, which digitizes the data and transfers to any Bluetooth accessible smart device to operate further to estimate and for the recording purpose. For the lossless data communication from the smart devices, it can be achieved by Bluetooth devices. In the sophisticated systems like mobile robots and manipulators consist of different types of sensors, which assist in accurate operation. Any error in the operation of such systems can be detected by Bluetooth platform connected to automation systems without disturbing its normal operation. This helps in detecting the cause of fault to maintenance personnel [6]. Larry W. Juang has proposed a battery monitoring system to estimate the state of charge, state of function, and state of health information conducted for the user utility. In the proposed methodology, internal states of battery measurements are performed externally for monitoring current and voltage remaining state. Power capability changes are detected by reduction in internal voltage and increase in impedance as the state of charge increases. Coulomb counting method is used to estimate the state of charge. State of function is determined by finding the impedance and open-circuit voltage of the battery as the minimum terminal voltage as the threshold for the battery monitoring system performance. To determine the state of health of the battery, relative impedance between original and present battery is compared from which wear and aging of the battery can be estimated [7].

3 Proposed Methodology

In the proposed wireless battery monitoring system for obstacle avoidance robot, the main constituents are hardware and software parts. The hardware consists of a Wireless Battery Monitoring System and Obstacle Avoidance Robot. The software programs are written using embedded C on Arduino-integrated development environment for robot operation and data acquisition from sensors connected to the battery. Parallax Data Acquisition (PLX-DAQ) is a software programming tool used in the PC for data analysis.

3.1 *Wireless Battery Monitoring System*

The battery monitoring system using wireless communication consists of lead–acid battery of 12 V, microcontroller ATMEGA 328, voltage divider circuit, current sensor, Bluetooth module, and laptop PC. Figure 1 depicts the block diagram of battery monitoring system using a wireless communication system. The flow of current is detected using Hall Effect sensor connected in series to the battery as shown in Fig. 1. The output voltage of the battery is detected using the voltage divider circuit by connecting the voltage divider circuit input terminals to battery terminals; output of the circuit is connected to the microcontroller. The continuously varying voltage and current values of the battery are converted into the required format, and then sent through serial port of microcontroller to Bluetooth module. The serial port protocol consists of HC-05 Bluetooth module designed for connecting serial wireless setup. The Parallax Data Acquisition (PLX-DAQ) software tool includes for Microsoft Excel in PC, which acquires data from microcontroller interfaced to it and stores the data in the Excel sheet as they arrive. PLX-DAQ has the property of plotting the graph as the data arrives in real time using Microsoft Excel.

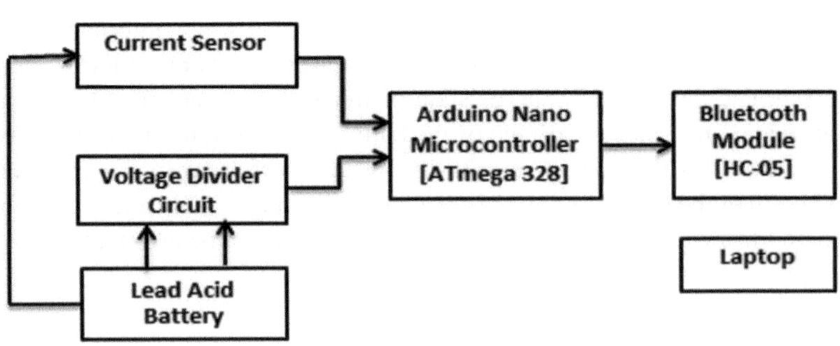

Fig. 1 Wireless battery voltage and current logging system

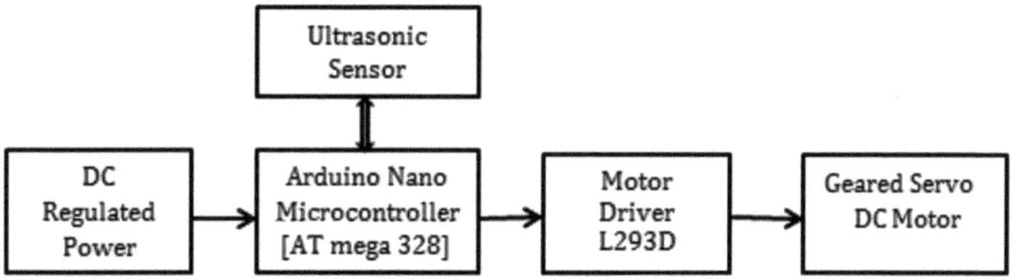

Fig. 2 Interconnection of obstacle avoidance robot

3.2 Obstacle Avoidance Robot

In the current work, obstacle avoidance robot is designed using ultrasonic sensor. The unexpected obstacles are avoided by getting collided by obstacle avoidance mobile robot autonomously. Block diagram of obstacle avoidance robot is shown in Fig. 2. The robot made using Arduino uses ultrasonic range sensor to avoid collisions. The robot is made of L293D interface circuit, Arduino board, and geared servo DC Motor. The robot controlling devices are connected to the board made using Arduino. The AT mega 328 microcontroller transmits signals to the L293D motor driver interfacing board, which controls the geared servo DC Motor.

4 Software Details

The software platforms used to evolve wireless monitoring system for battery are Arduino IDE and PLX-DAQ programming tools. The code is written using Arduino IDE and uploaded to the Arduino board. For Microsoft Excel, the parallax data acquisition is the added extra software feature. The parallax data acquisition software tool has the feature for analysis of collected data from sensors by using spreadsheeting. In Fig. 3, the flow diagram for programming wireless data acquisition system for battery monitoring system is depicted. By initializing the Bluetooth, port data can be received in the laptop. The parallax tool process the received data to record it in the Excel sheet.

5 Experimental Results

The continuously discharging voltage and current values of the lead–acid battery is measured using a voltage divider circuit and Hall Effect current sensor, respectively. Then it is converted into digital format and sent through serial port of microcontroller to Bluetooth communication module. The serial port protocol consists of HC-

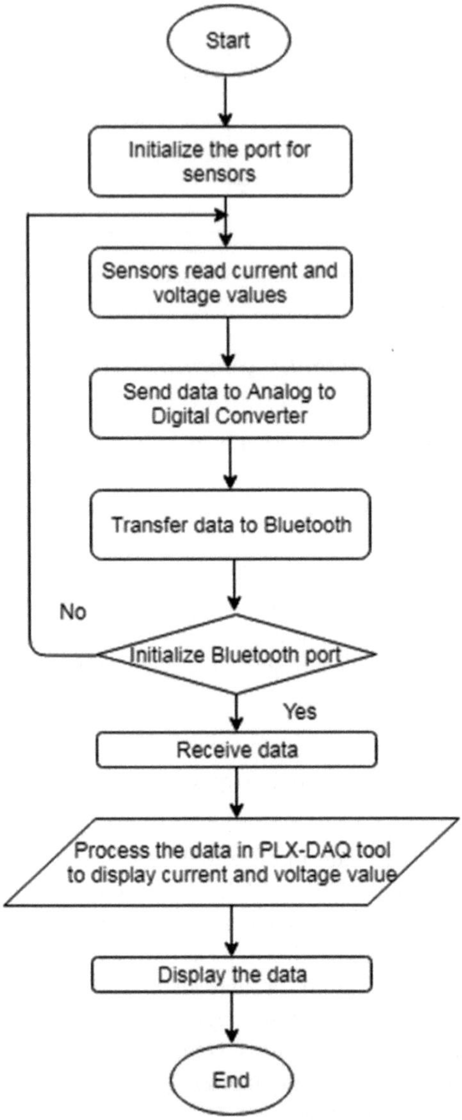

Fig. 3 Flowchart for wireless battery monitoring system

05 Bluetooth module designed for connecting serial wireless setup. The lead–acid battery of obstacle avoidance robot has been monitored for the discharging voltage and current by wireless monitoring system and tabulated the data into columns of the Microsoft Excel sheet in laptop.

In Table 1, it shows the discharging voltage and current values of the lead–acid battery transferred to Excel sheet. The graphical programming environment of PLX-DAQ software tool presents the display of discharging voltage and current waveforms

Table 1 Discharging voltage and current values of the battery

Sl. no.	Time (PM)	Difference time (s)	Voltage (V)	Current (A)
1	4:47:50	0	12	0.211
2	4:47:52	2.65625	11.75	0.106
3	4:47:54	4.65625	12	0.132
4	4:47:56	6.675781	11.75	−0.264
5	4:47:59	8.691406	12	−0.079
6	4:48:00	9.707031	12	0.079
7	4:48:02	12.67578	11.75	0.053
8	4:48:04	14.67578	11.75	0.238
9	4:48:07	16.69141	11.75	0.053
10	4:48:09	18.69141	12	0.132
11	4:48:11	20.72266	11.75	−0.158
12	4:48:12	21.73828	11.75	−0.026
13	4:48:17	26.74219	12	0.158
14	4:48:20	29.75781	11.75	0.053
15	4:48:23	32.72656	12	0.158
16	4:48:25	34.77344	12	0.211
17	4:48:27	36.86719	11.75	0.053
18	4:48:28	37.86719	12	−0.053
19	4:48:31	40.75781	11.75	−0.053
20	4:48:33	42.80469	12	0.132

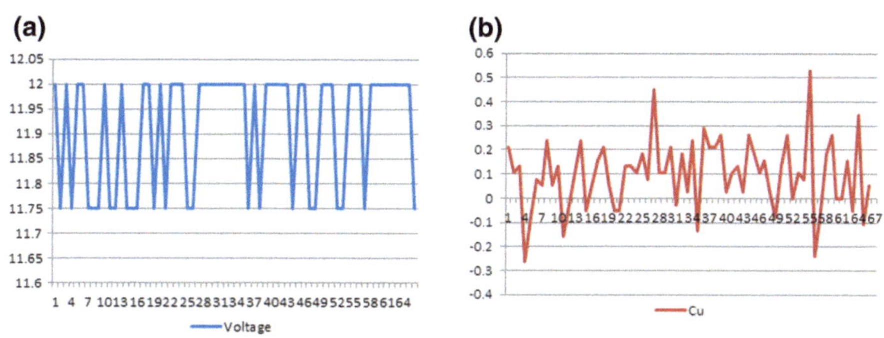

Fig. 4 **a** Graphical display of discharging voltage in volts with respect to time. **b** Graphical display of discharging current in amperes with respect to time

of the battery as shown in Fig. 4a, b. From the table and graph, it can be observed that voltage and current values are recorded and plotted at regular intervals of 2 s. The monitoring system notifies to the user on the PC if the battery voltage reaches a minimum threshold value of 10.5 V.

6 Conclusion

The wireless battery monitoring system provides the valuable and real-time discharging behavior of lead–acid battery of obstacle avoidance robot. The parameters of interest are voltage and current output of a discharging battery. The module of the Bluetooth is designed for wireless serial port connection setup. The laptop is interfaced easily to Bluetooth module for serial communication. In the experimental result shown, the system successfully measured voltage and current of battery and transmitted to Microsoft Excel on a computer. The PLX-DAQ software programming tool is used to establish an easy communication between Microsoft Excel on a Windows Computer and any device that supports serial port protocol.

References

1. Oka Danil Saputra SC, Kim YH, Shin SY (2015) Remote monitoring of lead-acid battery based on WLAN
2. Rauniyar A, Irfan M, Saputra OD, Kim JW, Lee AR, Jang JM, Shin SY (2017) Design and development of a real-time monitoring system for multiple lead-acid batteries based on internet of things. Futur Internet 9(3):28
3. Verma V, Tellapati R, Bayya M, Rao UM (2013) LabVIEW-based battery monitoring system with effects of temperature on lead-acid battery. Int J Enhanc Res Sci Technol Eng 2:6–10
4. Lee CY, Peng HC, Lee SJ, Hung I, Hsieh CT, Chiou CS, … Huang YP (2015) A flexible three-in-one microsensor for real-time monitoring of internal temperature, voltage and current of lithium batteries. Sensors 15(5):11485–11498
5. Jamaluddin A, Perdana FA, Supriyanto A, Purwanto A, Nizam M (2014, November) Development of wireless battery monitoring for electric vehicle. In: 2014 international conference on electrical engineering and computer science (ICEECS). IEEE, pp 147–151
6. Tašner T, Les K, Lovrec D (2013) Bluetooth platform for wireless measurements using industrial sensors. Int J Adv Rob Syst 10(1):75
7. Juang LW (2010) Online battery monitoring for state-of-charge and power capability prediction. University of Wisconsin Madison

Development of a GUI to Detect Glaucomatic Diseases Using Very Deep CNNs

G. Pavithra, T. C. Manjunath and T. N. Kesar

Abstract One of the deadliest diseases in human beings is the glaucoma, which is the second largest disease in the world, which leads to the loss of vision in the human eye, thus making the life of human miserable and the whole world would be dark without vision. Recently, (DL) Deep Learning is playing a lot of important role in the image processing applications. This DL can be clubbed with CNNs (Convolution Artificial Neural Networks) along with a hardware Raspberry Pi and the hybrid combination of the threesome could be used for the automated detection of the glaucomatic case in the disease-affected human beings in the eyes. In this write-up, the previously mentioned hybrid threesome is being used and developed for the glaucoma detection. The DL frameworks (CNN + ANN + MATLAB) can be used as a hierarchical representation of the fundus images to distinguish b/w glaucoma and non-glaucomatic images for the disease detections. The model is trained with standard datasets available on the net. The VGG19 architecture is used with transfer learning to achieve high accuracy. A graphical user interface is used to diagnose the condition of test images and give a graphical analysis of the patients. The entire program is run on a Raspberry Pi 3B with a 5" LCD touch screen as a stand-alone device with the power input.

Keywords Convolutional neural network · Computer vision · MNIST · Keras · Python · Classifier · Simulation · Results · ANN

G. Pavithra
VTU Regional Resource Centre, Belagavi, Karnataka, India

T. C. Manjunath (✉)
Electronics Communication Engineering Department, Dayananda Sagara College of Engineering, Bangalore, Karnataka, India
e-mail: dr.manjunath.phd@ieee.org

T. N. Kesar
ECE Department, DSCE, Bangalore, Karnataka, India

C. Shreesha and R. D. Gudi (eds.), *Control Instrumentation Systems*,
Lecture Notes in Electrical Engineering 581,
https://doi.org/10.1007/978-981-13-9419-5_13

1 Introduction

The eye disease, glaucoma is a gathering of eye ailments which result in the deterioration of the optic nerve connected to the brain, thus resulting in the loss of eye vision in both the human eye (both). As per the WHO standards, glaucoma is the second deadliest disease for visual impairment (vision loss) and it is in charge of roughly 5.2 million instances of eye disorders (15% of the aggregate weight of world visual impairment blindness) [1] and will take the tally to around 60 million individuals by the year 2030 [2]. It happens all the more ordinarily among more seasoned individuals from the newborn to the aged ones. Glaucoma has been known as the "quiet thief-criminal of sight" on the grounds that the loss of vision, as a rule, happens gradually over an extensive stretch of time.

Databases [3–10] are being used to train the model as inputs to the proposed algorithm. There is a frontend and a backend to the script. The Keras library for the implementation of the CNN is being used in the work considered. The main aim of Keras is modularity, a way of architecting the layers. Hence, the Keras deep learning library for implementing the proposed architecture is being used. Then, a method called as transfer learning where the weights of the VGG19 architecture are downloaded and are being used to create the model as the image dataset is not big enough are being used in the work. There are 550 images used for training and the validation is being done on 15% of them. The glaucomatous and the healthy images that are used for training are images taken by state-of-the-art fundus cameras. The first five layers of the VGG19 model are frozen and the final fully connected layers are customized to gain high accuracy. This model is compiled with loss function—categorical cross entropy and optimizer—stochastic gradient descent. A method called data augmentation is used on the input images and the model is generated. A checkpoint method is used to save the model when the accuracy is the highest. The model is compiled and fit on the laptop and is moved into the Raspberry Pi 3B.

The frontend of the script is run on the Raspberry Pi which loads the model. A built-in Python package called Tkinter is used for the graphical interface. A prediction algorithm is written in the frontend along with numerous other functionalities. The prognosis can be done on all the images of the test folder in an instant and there is no calculation required. A graphical representation of the diagnosis can also be shown. It creates a CSV (comma-separated values) file of all the test subjects with their diagnosis report. A preview option is also placed which can be used to view the fundus image in a dialog box.

The entire process of predicting the outcome is done on the Raspberry Pi which acts as a stand-alone device with the power input. The LCD 5-inch screen is placed on the module in order to view the result. The device can be carried anywhere with the model and can be used on different images. The glaucomatous and the healthy images can be diagnosed effortlessly.

2 Convolutional Neural Networks

A convolutional neural system (CNN, or ConvNet) is a class of profound, feed-forward fake neural systems that have effectively been connected to investigating visual symbolism. CNNs were enlivened by natural procedures in that the connectivity design between neurons looks like the association of the creature visual cortex. Individual cortical neurons react to boosts just in a limited district of the visual field known as the responsive field. The responsive fields of various neurons halfway cover with the end goal that they cover the whole visual field. Convolutional Neural Networks are fundamentally the same as common Neural Networks: they are comprised of neurons that have learnable weights and inclinations. Every neuron gets a few information sources, plays out a speck item and alternatively tails it with a non-linearity. The entire system still communicates a solitary differentiable score function: from the crude picture pixels towards one side to class scores at the other. Despite everything, they have a misfortune work (e.g. SVM/Softmax) on the last (completely associated) layer [11]. A typical convolution neural network (CNN) is shown in Fig. 1 [12], which is an advanced version of the high level artificial neural network (ANN) shown in Fig. 2, which consists of an i/p layer, no. of hidden layers and an o/p layer.

The reason behind choosing a very high-performance algorithm like CNN on a small data set of say 110 images is to get a very high accuracy, which could not be obtained in the other proposed algorithms (earlier works done by this article authors [13–17, 18–20]). It has to be noted that small data set is not being considered for the simulation purposes, similar to 110 images in 1 dataset, similar 10 datasets [3–10]

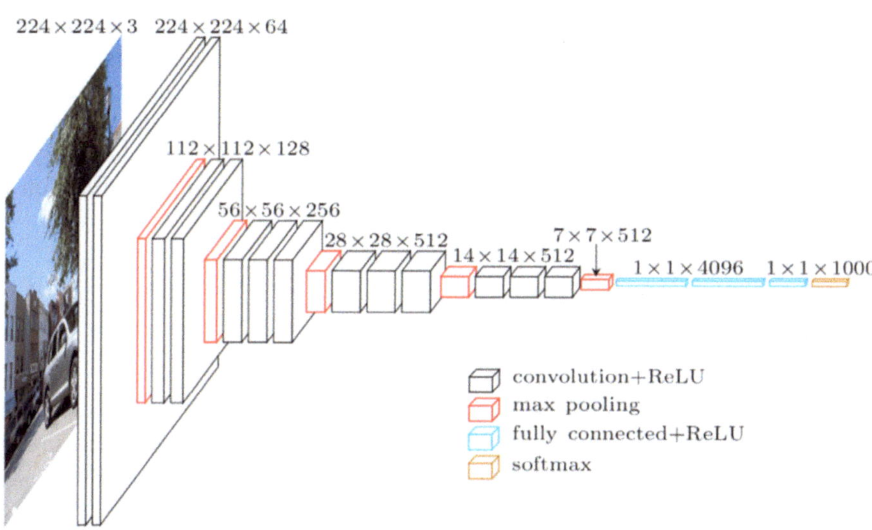

Fig. 1 Architecture of a deep CNN. *Source* [12]

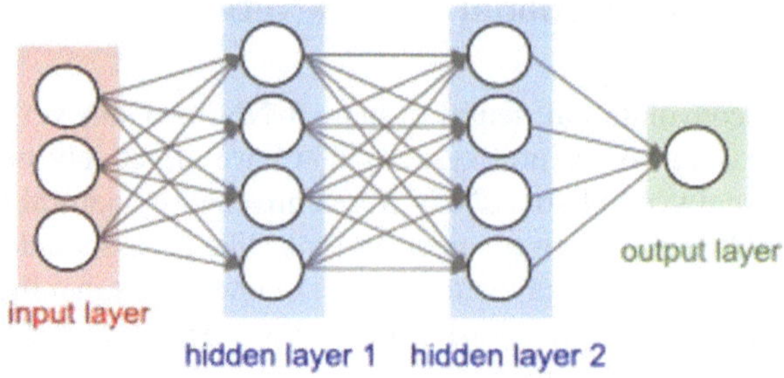

Fig. 2 Typical structural layout of an Artificial Neural Net (ANN). *Source* [22]

are being considered as a result of which 1100 images are being considered for the simulation purposes. This is done so that any input image is given from any database, the proposed algorithm works well with high accuracy.

3 Large-Scale Image Recognition Problem Identification Using Very Much Large Deep Convolution Neural Networks

The VGG19 investigates the effect of the convolutional neural networks depth on its accuracy in the large-scale image recognition setting. The principle commitment is a careful assessment of systems of expanding profundity utilizing an engineering with little (3 × 3) convolution channels, which demonstrates that a noteworthy change on the earlier craftsmanship arrangements can be accomplished by pushing the profundity to 16–19 weight layers. These discoveries were the premise of ImageNet challenge 2014 accommodation, where the group anchored the first place and the second place in the localization and characterization tracks individually. They additionally demonstrate that portrayals sum up well to different datasets, where they accomplish best in class results. The VGG best-performing religious circle models are made freely accessible to encourage additionally explore on the utilization of profound visual portrayals in PC vision.

4 Methodology

The dataset images are classified as glaucoma and no-glaucoma images and is placed in a separate folder for training. Images are fed to the neural network by Keras API—flow_from_directory where all images are resized to 256 width and 256 heights.

5 Transfer Learning VGG19 Architecture

This type of learning is an m/c learning methodology wherein a model that is going to be developed for a particular task is going to be re-used as the point of starting for a trained model which can be used for a second task. It takes a trained convolutional neural network and leverages the features that are extracted by that network on an input image and then applies these features for another task. So, eventually, training of a whole big network for all the the feature machine learning problems using AI concepts definitely needs for improving the accuracy.

6 Prediction

A Python package called Tkinter is used to provide graphical interface. Underwood a prediction algorithm is executed. The trained model which was saved in a .h5 format is loaded using Keras load function, The test image is taken as an input, resized to default ratio and a function model predict generates output predictions for the input sample and classifies the image as either with glaucoma or no glaucoma and show the output.

Prediction Classifiers—In the work considered, CNN is used as prediction classifiers for predicting whether the image is glaucomatic or not and the coding is done in the Python environment. Her, the prediction classifiers is defined as the process of predicting the class of given data points (yes or no). It predicts the class label correctly and the accuracy of the predictor refers to how well a given predictor can guess the value of predicted attribute for a new data and thus refers to the computational cost in generating and using the classifier or predictor.

7 Block Diagram

The proposed block diagram is shown in Fig. 3. The program is divided into three different categories, frontend, Backend and the Display. The backend code is executed on a GPU-enabled laptop. The Drishti dataset is loaded and the training is done based on the VGG19 model.

Fine-tuning is done on the fully connected layers to improve accuracy. The model created is saved on the Raspberry Pi. The Raspberry Pi loads the model and the test image is fed into the device. Preprocessing is done on the test image before the diagnosis is done. The preprocessed image is given to the prediction algorithm which outputs 1 if healthy and 0 if glaucomatous. These two classes are outputted on the GUI run on the module. The touch screen outputs the diagnosis of the patient along with the CSV file.

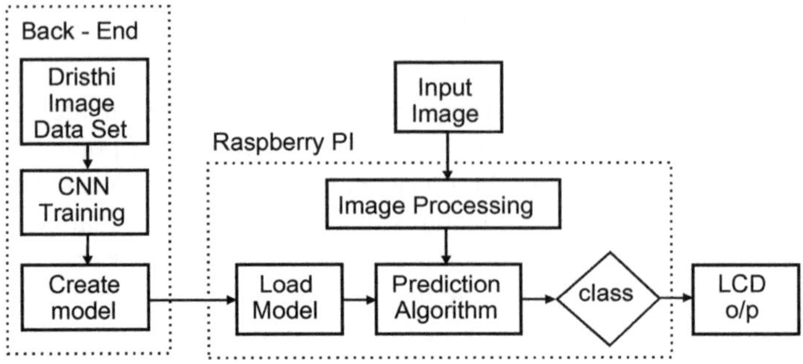

Fig. 3 Proposed block diagram

8 Advantages

Efficiency and Speed: Using CNN for image classification over traditional CDR calculation method is much more efficient as there are no requirements for manual feature extraction, which results in faster and more accurate outputs.
Mobility: The model has to be trained only once and can be used to for the classification of several images unlike in the former method where each image has to be passed through all stages of processing individually. Also, any embedded system can be used as a prediction device.
Diversity: The same model can be trained for other eye diseases, if the features to be extracted are similar like Diabetic retinopathy. This is far better than other methods as diabetic retinopathy doesn't involve CDR calculations.
Machine Learning: Based on the training dataset given, the model learns features and backpropagates to reduce the loss in each epoch. Data augmentation is also used to increase the number of features learnt from the limited dataset.

9 Result

After running the developed program, the simulation results are obtained as shown in Fig. 4. The trained model gives an accuracy of more than 98% which is loaded onto the Raspberry Pi 3B [21, 11, 13–17, 18–20]. The module uses the prediction algorithm and the images are classified. The *test folder-1* contains 110 images and the preview of each image can be seen along with the graphical representation. The GUI representation is shown above. Each test image can be run individually or together at once. It also gives a Pie chart of the diagnosis. Similarly, *test folder-2–10 are being considered as the inputs to algo* [3–10].

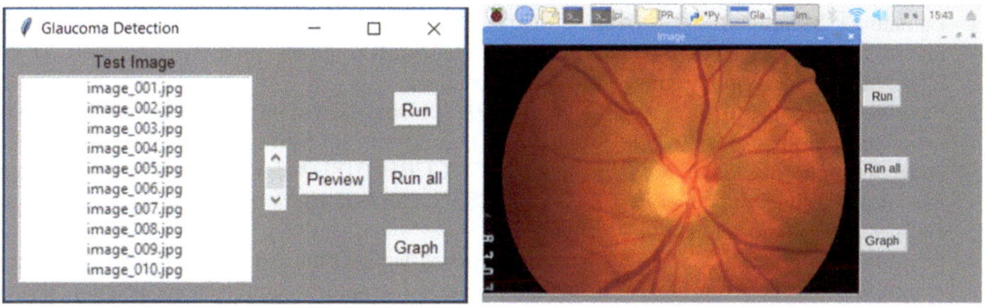

Fig. 4 GUI output of glaucoma detection

10 Conclusion

A deep convolutional network protocol has been presented in this research paper for detecting the glaucoma in the human beings. This D-CNN is able to get the main features of the glaucomatic persons, which characterizes the disease. The developed system is going to use a pre-training VGG19 CNN model along with the augmentation of the data, which is very much essential to predict the nature of the testing fundus eye images. The GUI built using Tkinter creates an interactive application that can be easily navigated. Each image can be tested individually or the entire dataset can be tested at once, resulting in a graphical representation of the results. In future work, a plan is being made to extend the CNN work to the study of D-L architecture which is based on the ANN-CNNs for the detection of multiple types of detection of ocular diseases. *It has to be noted that 10 sets of databases are being taken for simulation purposes and in the simulation results section, only one set of 110 images is being shown here for the sake of convenience as a result of which dataset considered is large and not small.*

References

1. Salam AA, Khalil T, Akram MU, Jameel A, Basit I (2016) Automated detection of glaucoma using structural and non-structural features. Springer Plus 5(1):1519
2. Chen Y, Xu, Kee Wong DW, Wong TY, Liu J (2015) Glaucoma detection based on deep convolution networks. In: 37th annual international conference of the IEEE engineering in medicine and biology society (EMBC), Milan, pp 715–718
3. https://www5.cs.fau.de/research/data/fundusimages/
4. http://cvit.iiit.ac.in/projects/mip/drishti-gs/mipdataset2/Home.php
5. https://www5.cs.fau.de/research/data/fundusimages/Opticdisc.org
6. http://www.optic-disc.org/library/normaldiscs/page7.html
7. https://www5.cs.fau.de/research/data/fundus-images/
8. http://www.ia.uned.es/~ejcarmona/DRIONS-DB.html
9. http://www.optic-disc.org/library/normal-discs/page7.html
10. http://www.isi.uu.nl/Research/Databases/DRIVE/download.php

11. Lamani D, Manjunath TC (2016, February) Diagnosis of glaucoma disease through image feature fractal dimension. PhD thesis, VTU, Belagavi, Karnataka
12. https://www.jeremyjordan.me/convnet-architectures/
13. Pavithra G, Manjunath TC (2018, December 19–20) Detection of primary glaucoma in humans using simple linear iterative clustering (SLIC) algorithm. In: Springer's international conference on computing networks, Big Data & IoT [ICCBI 2018], Vaigai College of Engineering, Madurai, Tamil Nadu, Paper id ICCBI-0108, ISSN: 2367–4512, Springer lecture notes on data engineering & communications technologies
14. Pavithra G, Manjunath TC (2018) Detection of primary glaucoma using fuzzy C mean clustering & morphological operators algorithm. In: Springer's international conference on computational vision & bio inspired computing (ICCVBIC 2018), paper id ICCVBIC-0155, ISSN: 1439-7358, Springer—lecture notes in computer science & engineering
15. Pavithra G, Manjunath TC (2017, May 19–20) Investigation of primary glaucoma by CDR in fundus images. In: 2nd IEEE international conference on recent trends in electronics, information & communication technologies (RTEICT-2017), SVCE, Bangalore, pp 1806–1812
16. Pavithra G, Manjunath TC (2017) Conceptual view of a smart tonopen for biomedical engineering applications. In: 2017 IEEE international conference on intelligent computing & control sciences (ICICCS-2017), Vaigai College of Engineering, Madurai, Tamil Nadu, pp 636–639
17. Pavithra G, Manjunath TC (2017, June 23–24) A review of the glaucoma detection using hardware based implementation using embedded systems. In: 2017 IEEE international conference on intelligent computing & control (I2C2–2017), Karpagam University, Coimbatore, pp 75–82
18. Manjunath TC (2015, September) Automated diagnose of neo-vascular glaucoma disease using advance image analysis technique. Int J Appl Info Syst (IJAS) 9(6):1–6
19. Manjunath TC (2016) A novel approach for diagnosis of glaucoma through optic nerve head (ONH) analysis using fractal dimension technique. Int J Mod Edn Comp Sci (IJMECS), 55–61
20. Manjunath TC (2016, January) A novel approach for diagnosis of glaucoma through optic nerve head (ONH) analysis using fractal dimension technique. Int J Mod Edn Comp Sci (IJMECS), 8(1):55–61
21. Manjunath TC (2017, May) Design of algorithms for diagnosis of primary glaucoma through estimation of CDR in different types of fundus images using IP techniques. Int J Innovat Res Info Secur (IJIRIS) 4(5):12–19
22. https://stackoverflow.com/questions/35345191/what-is-a-layer-in-a-neural-network
23. Sivaswamy J, Krishnadas KR, Joshi GD, Jain M, Ujjwal, SAT (2015) Drishti-GS: retinal image dataset for optic nerve head (ONH) segmentation. IEEE ISBI, Beijing
24. Orlando JI, Prokofyeva E, Del Fresno M, Blaschko MB (2017, January) Convolutional neural network transfer for automated glaucoma identification. In: 12th international symposium on medical information processing and analysis, vol 10160, p 101600U, International Society for Optics and Photonics
25. Almazroa A, Alodhayb S, Osman E, Ramadan E, Hummadi M, Dlaim M, Alkatee M, Raahemifar K, Lakshminarayanan V (2018, 6 March) Retinal fundus images for glaucoma analysis: the RIGA dataset. In: Proceedings of SPIE 10579, medical imaging 2018, imaging informatics for healthcare, research, and applications, 105790B
26. Simonyan K, Zisserman A (2015) Very deep convolutional networks for large-scale image recognition. arXiv:1409.1556v6 [cs.CV]
27. Pavithra G, Manjunath TC (2015, April) Different clinical parameters to diagnose glaucoma disease: a review. In: Proceedings of international journal of computer applications (IJCA), IF 3.546, ISSN 0975-8887. 116(23), pp 42–46
28. Sevastopolsky A (2017) Optic and cup segmentation methods for glaucoma detection with modification of U-net convolutional neural network. arXiv:1704.00979

MIX
Papier aus verantwortungsvollen Quellen
Paper from responsible sources
FSC® C105338

If you have any concerns about our products,
you can contact us on
ProductSafety@springernature.com

In case Publisher is established outside the EU,
the EU authorized representative is:
Springer Nature Customer Service Center GmbH
Europaplatz 3, 69115 Heidelberg, Germany

Printed by Libri Plureos GmbH
in Hamburg, Germany